S*pringer* **M***onographs* in **M***athematics*

For further volumes:
www.springer.com/series/3733

Anthony Tromba

A Theory of Branched Minimal Surfaces

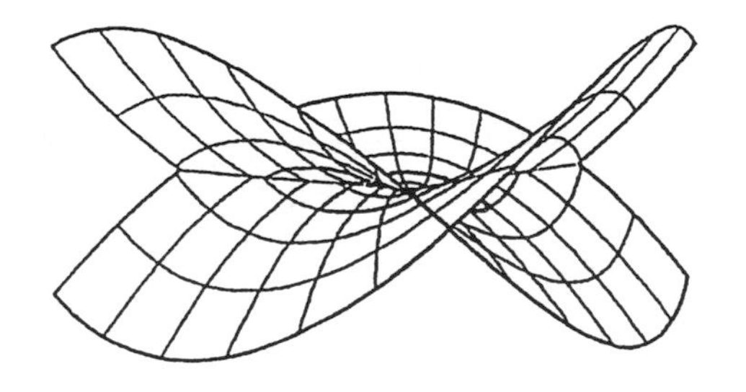

Springer

Anthony Tromba
Department of Mathematics
University of California at Santa Cruz
Santa Cruz, CA, USA

ISSN 1439-7382 Springer Monographs in Mathematics
ISBN 978-3-642-25619-6 e-ISBN 978-3-642-25620-2
DOI 10.1007/978-3-642-25620-2
Springer Heidelberg Dordrecht London New York

Library of Congress Control Number: 2011945164

Mathematics Subject Classification: 30B10, 49J50, 49Q05, 53A10, 58E12, 58C20

Printed on acid-free paper

Springer is part of Springer Science+Business Media (www.springer.com)

It is amazing how far you can come in life if you really know how to differentiate.

Jerrold E. Marsden

There are no problems in geometry requiring the calculation of a derivative greater than four.

S.S. Chern

A theorem isn't really proved until you can put it in a book.

Marty Golubitsky

To My Dear Friends

Michael Buchner
Stefan Hildebrandt
Jerry Marsden
Fritz Tomi

Preface

This book can be considered a continuation of *The Regularity of Minimal Surfaces* by Ulrich Dierkes, Stefan Hildebrandt and Anthony Tromba, Volume 340 of the *Grundlehren der Mathematischen Wissenchaften*.

The central theme is the study of branch points for minimal surfaces with the goal of providing a new approach to the elementary question of whether minima of area or energy must be immersed.

One of the main difficulties with the current theory of branch points is the transparency and sophistication of the proofs of the main theorems. For example, Osserman's original 1970 cut and paste proof, that absolute minima are free of interior branch points remains, for the most part, open only to experts. Furthermore, before the appearance of this volume, no complete proof has appeared in one place.

In the 1960's the development of global nonlinear analysis and the idea of doing calculus or analysis on infinite dimensional manifolds had created a great deal of excitement, especially through the pioneering work of Jim Eells, Dick Palais and Steve Smale.

The goal of this book is to develop entirely new and elementary methods, in the spirit of global analysis, to address this beautiful question via energy (Dirichlet's energy) as opposed to area. We will do something that rarely, if ever, has been done in the calculus of variations, namely calculate arbitrarily high orders of derivatives of energy. This method also applies to boundary branch points for minimal surfaces with smooth, but not analytic boundaries, a question that heretofore has not been addressed.

We wish to thank Stefan Hildebrandt for assisting with reworking part of the manuscript, but all errors are the sole responsibility of the author. A very special thanks must go to Daniel Wienholtz, whose brilliant insights led to the resolution of the boundary case for smooth curves, and finally to Fritz Tomi, who worked on the exceptional branch point case and pointed out potential difficulties in applying these methods to this case.

My career owes a great debt to all of these wonderful mathematicians.

Santa Cruz, USA
August 2011

Anthony Tromba

Contents

Chapter 1
Introduction

The classical problem of Plateau, although by far not the oldest problem in the Calculus of Variations, is certainly one of the best known. The mathematical formulation of the problem of finding a least area surface of the topological type of the disk spanning a closed contour goes back to Weierstrass. In particular, Weierstrass formulated the existence of the solution of the least area problem as a solution to a system of non-linear partial differential equations:

Set

$$B := \{w \in \mathbb{C} : |w| < 1\}$$

and

$$C := \{w \in \mathbb{C} : |w| = 1\} = \partial B.$$

A *closed Jordan curve* Γ *in* $\mathbb{R}$ is a subset of $\mathbb{R}^3$ which is homeomorphic to ∂B.

Given a closed Jordan curve Γ in $\mathbb{R}^3$ we say that $X : \overline{B} \to \mathbb{R}^3$ is a solution of Plateau's problem for the boundary contour Γ (or: a minimal surface spanned in Γ) if it fulfils the following three conditions:

(i) $X \in C^0(\overline{B}, \mathbb{R}^3) \cap C^2(B, \mathbb{R}^3)$;
(ii) The surface X satisfies in B the equations

$$\Delta X = 0 \tag{1.1}$$

$$|X_u|^2 = |X_v|^2, \quad \langle X_u, X_v \rangle = 0; \tag{1.2}$$

(iii) The restriction $X|C$ of X to the boundary C of the parameter domain B is a homeomorphism of C onto Γ.

From the classical point of view, one of the difficulties in minimizing the area functional

$$A_B(X) = \int_B |X_u \wedge X_v| \, du \, dv$$

is that among all those surfaces X satisfying (iii) A is invariant under the action of the infinite dimensional diffeomorphism group of B. By replacing area by energy one reduces the symmetry group to the finite dimensional conformal group of

A. Tromba, *A Theory of Branched Minimal Surfaces*,
Springer Monographs in Mathematics,
DOI 10.1007/978-3-642-25620-2_1, © Springer-Verlag Berlin Heidelberg 2012

the disk. Miraculously, the absolute minima of area and energy are the same. The Weierstrass equations (1.1) and (1.2) are then the variational equations of Dirichlet's energy.

The problem of the existence of a minimum of area spanning Γ remained open for a half a century until it was solved by Jesse Douglas (1931) and Tibor Radó (1930). For all his work on the Plateau problem, Douglas was awarded one of the first two Fields Medals of Mathematics (shared with Lars Ahlfors) at the International Congress of Mathematicians in Oslo in 1936.

Jesse Douglas (1897–1965)

Given the fact that the absolute minima of area and energy are the same, we can formulate the classical problem of Plateau as follows:

Given a closed Jordan curve Γ in $\mathbb{R}^3$, a mapping $X : B \to \mathbb{R}^3$ is said to be of class $\mathcal{C}(\Gamma)$ if $X \in H_2^1(B, \mathbb{R}^3)$, and if its trace $X|C$ can be represented by a weakly monotonic, continuous mapping $\varphi : C \to \Gamma$ of C onto Γ (i.e., every $L_2(C)$-representative of $X|C$ coincides with φ except for a subset of zero 1-dimensional Hausdorff measure).

Let

$$D(X) = D_B(X) := \frac{1}{2} \int_B (|X_u|^2 + |X_v|^2)\, du\, dv \tag{1.3}$$

be the Dirichlet integral of a mapping $X \in H_2^1(B, \mathbb{R}^3)$. Then we define the variational problem $\mathcal{P}(\Gamma)$ associated with Plateau's problem for the curve Γ as the following task:

Minimize Dirichlet's integral $D(X)$, *defined by* (1.3), *in the class* $\mathcal{C}(\Gamma)$.

In other words, setting

$$e(\Gamma) := \inf\{D(X) : X \in \mathcal{C}(\Gamma)\}, \tag{1.4}$$

we are to find a surface $X \in \mathcal{C}(\Gamma)$ such that

$$D(X) = e(\Gamma) \tag{1.5}$$

is satisfied.

In his solution, Douglas minimized an energy essentially equivalent to Dirichlet's energy, which later proved to be a very powerful method for dealing with minimal surfaces of arbitrary topological type and connectivity.

Almost from the beginning, the question arose as to whether the absolute minimizers were immersed or not. A point p where X is not immersed, i.e.

$$X_u(p) = X_v(p) = 0$$

is called a *branch point*. It follows easily that interior branch points are isolated. In 1932 Douglas [1] and in 1942 Courant [1] thought that they had found absolute minizers which had branch points. We should note here that from the early 1930s until his death in 1972 Courant worked on and popularized the field of minimal surfaces.

The example of Douglas was refuted in 1933 by Radó while Courant's example survived until the pioneering work of Robert Osserman in 1970, and then of Gulliver–Osserman and Royden in 1973.

In his now classic paper, Osserman constructed a discontinuous parameter transformation allowing a reparametrization of a minimal surface in a vicinity of an interior branch point, such that the area of the surface can be reduced. He had to distinguish between *true* and *false branch points* (the latter are those which have a neighbourhood whose image is still an embedded surface), but in his proof he overlooked some difficulties appearing for false branch points. In 1973, both H.W. Alt [1] and R. Gulliver [2] independently extended Osserman's line of argument to surfaces which are *absolute* minimizers of prescribed mean curvature with least energy and also treated the case of false branch points. The joint work of Gulliver, Osserman and Royden [1] in 1973 proved that *all* minimal surfaces bounded by rectifiable Jordan curves do not have any false branch points, even if they do not minimize the Dirichlet energy.

This difficult work has remained open mostly to experts in the field. For more historical comments, see the Scholia (Chap. 9).

In this book we give proof of the fact that in $\mathbb{R}^3$ any solution of Plateau's problem which is a *relative* minimizer of Dirichlet's integral D or, equivalently, the area functional A, is an immersion in the sense that it has no interior or (with mild assumptions) boundary branch points. This fact can easily be proved for planar boundaries (Dierkes, Hildebrandt and Sauvigny [1]), while the corresponding result in $\mathbb{R}^n$ is false for $n \geq 4$ according to a famous example of Federer. Therefore it remains to prove the assertion for a *nonplanar* boundary curve Γ in $\mathbb{R}^3$. The proof given here is based on the observation that one can compute any higher derivative of Dirichlet's integral in the direction of so-called *(interior) forced Jacobi fields*, using methods of complex analysis such as power series expansions and Cauchy's integral theorem as well as the residue theorem. These Jacobi fields lie in the kernel of the second variation of D; they also play a fundamental role in the index theory and the Morse theory of minimal surfaces. So, in a very strong sense, this book is about energy and the fact that it can be reduced in the presence of an interior or boundary branch point. This is in the spirit of Douglas' original approach to the Plateau problem. Since area is less than or equal to energy, reducing energy means that you can also reduce area. In this connection we must mention the work of Beeson [1].

Although the computations in this book are sometimes tedious, they are simple in principle. The main analytical idea is to find, using function theory, paths so that the calculation of higher order derivatives of Dirichlet's energy, through the use of Cauchy's integral theorem, along these paths reduces to a few manageable terms. In a sense, we are doing calculus on infinite dimensional manifolds. In order to convey to the reader a feeling for the methods to be applied, we begin by calculating the first five derivatives of Dirichlet's integral in the direction of special types of forced Jacobi fields, thereby establishing that a relative D-minimizing solution of Plateau's problem cannot have certain kinds of interior branch points. These introductory calculations will be carried out in Chap. 2 as a warm up for the general case, together with an outline of the variational procedure to be used in the sequel. These calculations are made transparent by shifting the branch point that is studied into the origin, and by bringing the minimal surface into a *normal form* with respect to the branch point $w = 0$ with an *order* n. Then also the *index* m of this branch point can be defined, with $m > n$. Furthermore, $w = 0$ is called an *exceptional branch point* if there is an integer $\kappa > 1$ such that $m + 1 = \kappa(n + 1)$. This notion is related to that of the false branch point, but it is a weaker notion. It will turn out that it is particularly difficult to exclude that a relative minimizer of D can have an exceptional branch point at $w = 0$. In fact, we are only able to exclude exceptional branch points for weak relative minimizers of A in $\mathcal{C}(\Gamma)$. However, we do present conditions under which a minimal surface with an exceptional branch point cannot be a relative minimizer of D. In the non-exceptional case, one can "always" reduce energy (and area), and surprisingly the monotonicity of a minimal surface on the boundary plays no role in being able to do so.

In Chap. 2 it is described how the variations $\hat{Z}(t)$ of a minimal surface $\hat{X}$ are constructed by using interior forced Jacobi fields. This leads to the (rather weak) notion of a *weak minimizer of* D. Any absolute or weak relative minimizer of D in $\mathcal{C}(\Gamma)$ will be a weak D-minimizer, and the aim is to investigate whether such minimizers can have $w = 0$ as an interior branch point. This possibility is excluded if one can find an integer $L \geq 3$ and a variation $\hat{Z}(t)$ of $\hat{X}$, $|t| \ll 1$, such that $E(t) := D(\hat{Z}(t))$ satisfies

$$E^{(j)}(0) = 0 \quad \text{for } 1 \leq j \leq L - 1, \qquad E^{(L)}(0) < 0.$$

It will turn out that the existence of such an L depends on the order n and the index m of the branch point $w = 0$.

In our first chapter, this idea is studied by investigating the third, fourth and fifth derivatives of $E(t)$ at $t = 0$. Here one meets fairly simple cases for testing the technique demonstrating its efficacy. Furthermore, the difficulties are exhibited that will come up generally.

The first case of a general nature is treated in Chap. 3. Assuming that $n+1$ is even and $m + 1$ is odd (whence $w = 0$ is non-exceptional) it will be seen that $E^{(m+1)}(0)$ can be made negative while $E^{(j)}(0) = 0$ for $1 \leq j \leq m$, and so $\hat{X}$ cannot be a weak minimizer of D.

The general situation is studied in Chaps. 4 to 7. In Chap. 4 is shown that $w = 0$ cannot be a non-exceptional branch point of a weak relative minimizer of D. We derive simple formulae for the first non-vanishing derivatives of Dirichlet's energy and show that they can be made negative. Such a result is no longer true for an exceptional branch point $w = 0$, apart from some special cases. In Chaps. 5, 6 and 7 it is proved that a weak relative minimizer of A in $\mathcal{C}(\Gamma)$ cannot have exceptional interior branch points if Γ is a smooth closed Jordan curve in $\mathbb{R}^3$.

In Chap. 8 we study boundary branch points of a minimal surface $\hat{X}$ with a smooth boundary contour. In particular we first show that $\hat{X}$ cannot be a weak relative minimizer of D if it has a boundary branch point whose order n and index m satisfy the condition $2m - 2 < 3n$ (Wienholtz's theorem).

We then will show that if the torsion and curvature of Γ are both non-zero, then a priori $2m + 2 \leq 6(n + 1)$. As a consequence it follows that $\hat{X}$ is not a minimizer in the non-exceptional cases; i.e. $m + 1 \neq k(n + 1)$, $k = 2$ or 3. This is a partial resolution to boundary regularity for smooth contours. Considering only the Taylor expansion about a branch point, we then argue that the question of whether a minimal surface with an exceptional boundary branch point is or is not a minimum is not decidable.

In conclusion, if the boundary contour is C^∞ or more simply if a minimal surface $\hat{X}$ is C^∞ with a non-exceptional interior or boundary branch point, we can find a C^∞ surface Y which is C^∞ close to $\hat{X}$ having less energy and area. This is much stronger than what was previously known and indicates the power of using derivatives as opposed to cut and paste constructions.

In the Scholia (Chap. 9) we describe some of the history of the main results of this book. Finally, we note that some of the introductory material also appears in Dierkes, Hildebrandt and Tromba [1], but we include it for completeness. The author wishes to thank Stefan Hildebrandt for reworking the manuscript and for his encouragement, the Max Planck Institute in Leipzig for their support, Frau Birgit Dunkel for her excellent typing of the manuscript and finally my wife Inga without whose love and support this book could not have been written.

Chapter 2
Higher Order Derivatives of Dirichlet's Energy

2.1 First Five Variations of Dirichlet's Integral and Forced Jacobi Fields

In this chapter we take the point of view of Jesse Douglas and consider minimal surfaces as critical points of Dirichlet's integral within the class of harmonic surfaces $X : B \to \mathbb{R}^3$ that are continuous on the closure of the unit disk B and map $\partial B = S^1$ homeomorphically onto a closed Jordan curve Γ of $\mathbb{R}^3$. It will be assumed that Γ *is smooth of class* C^∞ *and nonplanar*. Then any minimal surface bounded by Γ will be a nonplanar surface of class $C^\infty(\overline{B}, \mathbb{R}^3)$, and so we shall be allowed to take directional derivatives (i.e. "variations") of any order of the Dirichlet integral along an arbitrary C^∞-smooth path through the minimal surface.

The first goal is to develop a *technique* which enables us to compute variations of any order of Dirichlet's integral, D, at an arbitrary minimal surface bounded by Γ, using complex analysis in the form of Cauchy's integral theorem. This will be achieved by varying a given minimal surface via a one-parameter family of admissible harmonic mappings. Such harmonic variations will be generated by varying the boundary values of a given minimal surface in an admissible way and then extending the varied boundary values harmonically into B. From this point of view the admissible boundary maps $\partial B = S^1 \to \Gamma$ are the primary objects while their harmonic extensions $\overline{B} \to \mathbb{R}^3$ are of secondary nature. This calls for a change of notation: An admissible boundary map will be denoted by $X : \partial B \to \Gamma$, whereas $\hat{X}$ is the uniquely determined harmonic extension of X into B; i.e. $\hat{X} \in C^0(\overline{B}, \mathbb{R}^3) \cap C^2(B, \mathbb{R}^3)$ is the solution of

$$\triangle \hat{X} = 0 \quad \text{in } B, \qquad \hat{X}(w) = X(w) \quad \text{for } w \in \partial B.$$

Instead of $\hat{X}$ we will occasionally write HX or $H(X)$ for this extension, and

$$D(\hat{X}) := \frac{1}{2} \int_B \nabla \hat{X} \cdot \nabla \hat{X} \, du \, dv$$

is its Dirichlet integral.

In the sequel the main idea is to vary the boundary values X of a given minimal surface $\hat{X}$ in the direction of a so-called *forced Jacobi field*, as this restriction will

A. Tromba, *A Theory of Branched Minimal Surfaces*,
Springer Monographs in Mathematics,
DOI 10.1007/978-3-642-25620-2_2,

enable us to evaluate the variations of D at X by means of Cauchy's integral theorem. In order to explain what forced Jacobi fields are we first collect a few useful formulae.

Let us begin with an arbitrary mapping $X \in C^\infty(\partial B, \mathbb{R}^n)$ and its harmonic extension $\hat{X} \in C^\infty(\overline{B}, \mathbb{R}^3)$. Then $\hat{X}$ is of the form

$$\hat{X}(w) = \operatorname{Re} f(w) \tag{2.1}$$

where f is holomorphic on B and can be written as

$$f = \hat{X} + i\hat{X}^* \quad \text{with } \hat{X}_u = \hat{X}_v^* \text{ and } \hat{X}_v = -\hat{X}_u^*. \tag{2.2}$$

We also note that

$$f'(w) = 2\hat{X}_w(w) = \hat{X}_u(w) - i\hat{X}_v(w) \quad \text{in } B. \tag{2.3}$$

Conversely, *if f is holomorphic in B and $\hat{X} = \operatorname{Re} f$ then f' and $\hat{X}_w$ are related by the formula $f' = 2\hat{X}_w$*; in particular, *$\hat{X}_w$ is holomorphic in B*. This simple, but basic fact will be used repeatedly in later computations.

Let us introduce polar coordinates r, θ about the origin by $w = re^{i\theta}$, and set $\hat{Y}(r,\theta) = \hat{X}(re^{i\theta})$. Then a straightforward computation yields

$$iw\hat{X}_w(w)\Big|_{w=e^{i\theta}} = \frac{1}{2}\left[\hat{Y}_\theta(1,\theta) + i\hat{Y}_r(1,\theta)\right] \tag{2.4}$$

whence

$$2\operatorname{Re}\left\{iw\hat{X}_w(w)\right\}\Big|_{w=e^{i\theta}} = \hat{Y}_\theta(1,\theta) = \frac{\partial}{\partial\theta}X(e^{i\theta}) = Y_\theta(\theta) \tag{2.5}$$

since

$$\hat{Y}(1,\theta) = \hat{X}(e^{i\theta}) = X(e^{i\theta}) =: Y(\theta).$$

If $X \in C^\infty(S^1, \mathbb{R}^3)$ maps S^1 homeomorphically onto Γ then $Y_\theta(\theta)$ is tangent to Γ at $Y(\theta)$, i.e. $Y_\theta(\theta) \in T_{Y(\theta)}\Gamma$, and so the left-hand side of (2.5) is tangent to Γ.

Consider now a continuous function $\tau : \overline{B} \to \mathbb{C}$ that is meromorphic in B with finitely many poles in B, and that is real on ∂B. Then τ can be extended to a meromorphic function on an open set Ω with $\overline{B} \subset \Omega$, and τ is holomorphic in a strip containing ∂B. It follows from (2.5) that

$$2\operatorname{Re}\left\{iw\hat{X}_w(w)\tau(w)\right\}\Big|_{w=e^{i\theta}} = \tau(e^{i\theta})Y_\theta(\theta) \in T_{Y(\theta)}\Gamma. \tag{2.6}$$

Suppose now that $\hat{X}$ is a minimal surface with finitely many branch points in $\overline{B}$. These points are the zeros of the function $F(w) := \hat{X}_w(w)$ which is of class C^∞ on $\overline{B}$ and holomorphic in B. If $\tau(w)$ has its poles at most at the (interior) zeros of the function $wF(w)$, and if the order of any pole does not exceed the order of the corresponding zero of $wF(w)$, then the function $K(w) := iw\hat{X}_w(w)\tau(w)$ is holomorphic in B and of class $C^\infty(\overline{B}, \mathbb{R}^3)$. We call $\hat{h} := \operatorname{Re} K$ an **inner forced Jacobi field** $\hat{h} : \overline{B} \to \mathbb{R}^3$ **at** $\hat{X}$ with the **generator** τ.

If one wants to study boundary branch points of $\hat{X}$ it will be useful to admit factors $\tau(w)$ which are meromorphic on $\overline{B}$, real on ∂B, with poles at most at the zeros of $wF(w)$, the pole orders not exceeding the orders of the associated zeros of $wF(w)$. Then

$$\hat{h} := \operatorname{Re} K \quad \text{with } K(w) := iwF(w)\tau(w),\ w \in \overline{B},\ F := \hat{X}_w, \tag{2.7}$$

is said to be a (general) **forced Jacobi field** $\hat{h} : \overline{B} \to \mathbb{R}^3$ at the minimal surface $\hat{X}$, and τ is called the **generator** of $\hat{h}$.

The boundary values $\hat{h}|_{S^1}$ of a forced Jacobi field $\hat{h}$ are given by

$$h(\theta) := \hat{h}(e^{i\theta}) = \operatorname{Re} K(e^{i\theta}) = \frac{1}{2}\tau(e^{i\theta})Y_\theta(\theta), \quad Y(\theta) := \hat{X}(e^{i\theta}). \tag{2.8}$$

Using the asymptotic expansion of $F(w) = X_w(w)$ at a branch point $w_0 \in \overline{B}$ having the order $\lambda \in \mathbb{N}$, we obtain the factorization

$$F(w) = (w - w_0)^\lambda G(w) \quad \text{with } G(w_0) \neq 0, \tag{2.9}$$

and, using Taylor's expansion in B or Taylor's formula on ∂B respectively, it follows that $G(w) = G(u, v)$ is a holomorphic function of w in B and a C^∞-function of $(u, v) \in \overline{B}$. It follows that *any forced Jacobi field* $\hat{h} : \overline{B} \to \mathbb{R}^3$ *is of class* $C^\infty(\overline{B}, \mathbb{R}^3)$ *and harmonic in* B.

Denote by $J(\hat{X})$ the linear space of forced Jacobi fields at $\hat{X}$, and let $J_0(\hat{X})$ be the linear subspace of inner forced Jacobi fields. The importance of $J(\hat{X})$ arises from the fact that *every forced Jacobi field* $\hat{h}$ *at* $\hat{X}$ *annihilates the second variation of* D, *i.e.*

$$\delta^2 D(\hat{X}, \hat{h}) = 0 \quad \text{for all } \hat{h} \in J(\hat{X}).$$

In the present section we only deal with inner forced Jacobi fields, and so we only prove the weaker statement (cf. Proposition 2.1):

$$\delta^2 D(\hat{X}, \hat{h}) = 0 \quad \text{for all } \hat{h} \in J_0(X).$$

The existence of forced Jacobi fields arises from the group of conformal automorphisms of $\overline{B}$ and from the presence of branch points; the more branch points $\hat{X}$ has, and the higher their orders are, the more Jacobi fields appear – this explains the adjective "forced". To see the first statement we consider one-parameter families of conformal automorphisms $\varphi(\cdot, t)$, $|t| < \epsilon$, $\epsilon > 0$ of $\overline{B}$ with

$$w \mapsto \varphi(w, t) = w + t\eta(w) + o(t) \quad \text{and} \quad \varphi(w, 0) = w, \dot{\varphi}(w, 0) = \eta(w). \tag{2.10}$$

Type I:

$$\varphi_1(w, t) = e^{i\alpha(t)}w$$

with $\alpha(t) \in \mathbb{R}$, $\alpha(0) = 0$, $\dot{\alpha}(0) = a$. Then $\varphi_1(w, t) = w + tiwa + o(t)$, and so

$$\eta_1(w) = iwa \quad \text{with } a \in \mathbb{R}.$$

Type II:

$$\varphi_2(w, t) := \frac{w + i\beta(t)}{1 - i\beta(t)w}$$

with $\beta(t) \in \mathbb{R}$, $\beta(0) = 0$, $\dot{\beta}(0) = b$.

Then $\varphi_2(w,t) = w + t\eta_2(w) + o(t)$ with $\eta_2(w) = ib + ibw^2$, and so

$$\eta_2(w) = iw\left(\frac{b}{w} + bw\right) \quad \text{with } b \in \mathbb{R}.$$

Type III:

$$\varphi_3(w,t) := \frac{w - \gamma(t)}{1 - \gamma(t)w}$$

with $\gamma(t) \in \mathbb{R}$, $\gamma(0) = 0$, $\dot{\gamma}(0) = c$.

Then $\varphi_3(w,t) = w + t\eta_3(w) + o(t)$ with $\eta_3(w) = -c + cw^2$, whence

$$\eta_3(w) = iw\left(\frac{ic}{w} - icw\right).$$

We set

$$\tau_1(w) := a, \qquad \tau_2(w) := b \cdot \left(\frac{1}{w} + w\right), \qquad \tau_3(w) := c \cdot \left(\frac{i}{w} - iw\right), \tag{2.11}$$

with arbitrary constants $a, b, c \in \mathbb{R}$. For $w = e^{i\theta} \in \partial B$ we have

$$\tau_1(w) = a, \qquad \tau_2(w) = 2b\cos\theta, \qquad \tau_3(w) = -2c\sin\theta,$$

and so τ_j, $j = 1, 2, 3$, are generators of the "special" forced Jacobi field $\hat{h}_j := \operatorname{Re} K_j$, defined by

$$K_j(w) := iwF(w)\tau_j(w), \quad w \in \overline{B}, \; F := \hat{X}_w, \tag{2.12}$$

which are inner forced Jacobi fields for any minimal surface $\hat{X}$ bounded by Γ. If we vary $\hat{X}$ by means of $\varphi = \varphi_1, \varphi_2, \varphi_3$ with $\alpha := \operatorname{Re}\varphi$, $\beta := \operatorname{Im}\varphi$, i.e. $\varphi(w,t) = \alpha(u,v,t) + i\beta(u,v,t)$, setting

$$\hat{Z}(w,t) := \hat{X}(\varphi(w,t)) = \hat{X}(\alpha(u,v,t), \beta(u,v,t)),$$

we obtain

$$\begin{aligned} \frac{d}{dt}\hat{Z} = \frac{d}{dt}\hat{X} \circ \varphi = \frac{d}{dt}\hat{X}(\alpha,\beta) &= \hat{X}_u(\alpha,\beta)\dot{\alpha} + \hat{X}_v(\alpha,\beta)\dot{\beta} \\ &= 2\operatorname{Re}\hat{X}_w(\varphi)\dot{\varphi}, \end{aligned}$$

and so

$$\left.\frac{d}{dt}\hat{Z}\right|_{t=0} = 2\operatorname{Re}\{\hat{X}_w\dot{\varphi}(0)\}.$$

For $\varphi = \varphi_j$ we have $\dot{\varphi}(0) = \eta_j$, hence

$$\left.\frac{d}{dt}\hat{Z}(w,t)\right|_{t=0} = 2\operatorname{Re}\{iw\hat{X}_w(w)\tau_j(w)\} = 2\hat{h}_j(w). \tag{2.13}$$

Let us now generate variations $\hat{Z}(t)$, $|t| \ll 1$, of a minimal surface $\hat{X}$ using any inner forced Jacobi field $\hat{h} \in J_0(\hat{X})$. We write $\hat{Z}(t) = \hat{Z}(\cdot,t)$ for the variation of $\hat{X}$ and $Z(t)$ for the variation of the boundary values X of $\hat{X}$, and start with the definition of $Z(t)$. Then $\hat{Z}(t)$ will be defined as the harmonic extension of $Z(t)$, i.e.

$$\hat{Z}(t) = H(Z(t)). \tag{2.14}$$

First we pick a smooth family $\gamma(t) = \gamma(\cdot, t)$, $|t| < \delta$, of smooth mappings $\gamma(t) : \mathbb{R} \to \mathbb{R}$ with $\gamma(0) = \mathrm{id}_{\mathbb{R}}$ which are "shift periodic" with the period 2π, i.e.

$$\gamma(\theta, 0) = \theta \quad \text{and} \quad \gamma(\theta + 2\pi, t) = \gamma(\theta, t) + 2\pi \quad \text{for } \theta \in \mathbb{R}. \tag{2.15}$$

Setting $\sigma(\theta, t) := \gamma(\theta, t) - \theta$ we obtain

$$\gamma(\theta, t) = \theta + \sigma(\theta, t) \quad \text{with } \sigma(\theta, 0) = 0 \text{ and } \sigma(\theta + 2\pi, t) = \sigma(\theta, t)$$

and

$$\gamma_\theta(\theta, t) = 1 + \sigma_\theta(\theta, t) = 1 + \sigma_{\theta t}(\theta, 0)t + o(t).$$

Choosing $\delta > 0$ sufficiently small it follows that

$$\gamma_\theta(\theta, t) > 0 \quad \text{for } (\theta, t) \in \mathbb{R} \times (-\delta, \delta).$$

Now we define the variation $\{Z(t)\}_{|t|<\delta}$ of X by

$$Z(e^{i\theta}, t) := X(e^{i\gamma(\theta,t)}) = \hat{X}(\cos\gamma(\theta, t), \sin\gamma(\theta, t)). \tag{2.16}$$

Then

$$\frac{\partial}{\partial t} Z(e^{i\theta}, t) = \left[-\hat{X}_u(e^{i\gamma(\theta,t)}) \sin\gamma(\theta, t) + \hat{X}_v(e^{i\gamma(\theta,t)}) \cos\gamma(\theta, t) \right] \gamma_t(\theta, t).$$

By (2.4) we have

$$ie^{i\theta} \hat{X}_w(e^{i\theta}) = \frac{1}{2} \left[X_\theta(\theta) + i\hat{X}_r(1, \theta) \right]$$

if we somewhat sloppily write $\hat{X}(r, \theta)$ for $\hat{X}(re^{i\theta})$ and $X(\theta)$ for $\hat{X}(1, \theta) = X(e^{i\theta})$. This leads to

$$-\hat{X}_u(e^{i\gamma(\theta,t)}) \sin\gamma(\theta, t) + \hat{X}_v(e^{i\gamma(\theta,t)}) \cos\gamma(\theta, t) = X_\theta(\gamma(\theta, t))$$

whence

$$\frac{\partial}{\partial t} Z(e^{i\theta}, t) = X_\theta(\gamma(\theta, t))\gamma_\theta(\theta, t) \cdot \frac{\gamma_t(\theta, t)}{\gamma_\theta(\theta, t)}.$$

On account of

$$Z(\theta, t) := Z(e^{i\theta}, t) = X(\gamma(\theta, t)) \tag{2.17}$$

we have

$$Z_\theta(\theta, t) = X_\theta(\gamma(t, \theta)) \cdot \gamma_\theta(\theta, t),$$

and so it follows that

$$\frac{\partial}{\partial t} Z(e^{i\theta}, t) = \frac{\partial}{\partial t} Z(\theta, t) = \frac{\partial}{\partial \theta} Z(\theta, t) \cdot \phi(\theta, t)$$

with

$$\phi(\theta, t) := \frac{\gamma_t(\theta, t)}{\gamma_\theta(\theta, t)}. \tag{2.18}$$

Defining the family $\{\phi(t)\}_{|t|<\delta}$ of 2π-periodic functions $\phi(t): \mathbb{R} \to \mathbb{R}$ by $\phi(t) := \phi(\cdot, t)$, we have

$$\frac{\partial}{\partial t} Z(t) = \phi(t) Z(t)_\theta =: h(t). \tag{2.19}$$

Now we consider the varied Dirichlet integral

$$E(t) := D(\hat{Z}(t)) = \frac{1}{2} \int_B \nabla \hat{Z}(t) \cdot \nabla \hat{Z}(t)\, du\, dv. \tag{2.20}$$

Then

$$\frac{d}{dt} E(t) = \int_B \nabla \hat{Z}(t) \cdot \nabla \frac{d}{dt} \hat{Z}(t)\, du\, dv.$$

Since the operations $\frac{d}{dt}$ and H commute, we have

$$\frac{d}{dt} \hat{Z}(t) = H\left(\frac{d}{dt} Z(t) \right)$$

and therefore

$$\frac{d}{dt} E(t) = \int_B \nabla \hat{Z}(t) \cdot \nabla H\left(\frac{d}{dt} Z(t) \right) du\, dv.$$

Since $\triangle \hat{Z}(t) = 0$, an integration by parts leads to

$$\frac{d}{dt} E(t) = \int_0^{2\pi} \frac{\partial}{\partial r} \hat{Z}(t) \cdot h(t)\, d\theta \quad \text{with } h(t) = \frac{\partial}{\partial t} Z(t). \tag{2.21}$$

For brevity we write in the following computations $\hat{Z}$ instead of $\hat{Z}(t)$. We have

$$w \hat{Z}_w = \frac{1}{2} (\hat{Z}_r - i \hat{Z}_\theta)$$

if we write $\hat{Z}(r, \theta)$ for $\hat{Z}(w)|_{w = re^{i\theta}}$, cf. (2.4), and also

$$dw = i w d\theta \quad \text{for } w = e^{i\theta} \in \partial B.$$

Then on ∂B:

$$\begin{aligned} w \hat{Z}_w \cdot \hat{Z}_w dw &= i (w \hat{Z}_w) \cdot (w \hat{Z}_w)\, d\theta \\ &= \frac{i}{4} (\hat{Z}_r - i \hat{Z}_\theta) \cdot (\hat{Z}_r - i \hat{Z}_\theta)\, d\theta \\ &= \left[\frac{1}{2} \hat{Z}_r \cdot \hat{Z}_\theta - \frac{i}{4} (\hat{Z}_r \cdot \hat{Z}_r - \hat{Z}_\theta \cdot \hat{Z}_\theta) \right] d\theta, \end{aligned}$$

and so

$$2\,\mathrm{Re}[w \hat{Z}_w \cdot \hat{Z}_w\, \phi\, dw] = \hat{Z}_r \cdot \hat{Z}_\theta\, \phi\, d\theta \quad \text{on } \partial B.$$

Furthermore, $\hat{Z}_\theta = Z_\theta$ on ∂B as well as $h = \phi Z_\theta$ (see (2.19)), and so (2.21) leads to the formula

$$\frac{d}{dt} E(t) = 2\,\mathrm{Re} \int_{S^1} w \hat{Z}(t)_w \cdot \hat{Z}(t)_w \phi(t)\, dw \tag{2.22}$$

where the closed curve S^1 is positively oriented. This formula will be the starting point for calculating all higher order derivatives $\frac{d^n}{dt^n}E(t)$ and, in particular, of $\frac{d^n}{dt^n}E(0) := \frac{d^n}{dt^n}E(t)\big|_{t=0}$. In order to evaluate the latter expressions for any n, it will be essential that we can choose $\phi(t)$ and any number of t-derivatives of $\phi(t)$ in an arbitrary way. This is indeed possible according to the following result:

Lemma 2.1 *By a suitable choice of $\gamma(\theta,t) = \theta + \sigma(\theta,t)$ with $\sigma \in C^\infty$ on $\mathbb{R} \times (-\delta,\delta)$, $\sigma(\theta,0) = 0$ and $\sigma(\theta+2\pi,t) = \sigma(\theta,t)$ we can ensure that the variation of the boundary values of the minimal surface $\hat{X}$, defined by $Z(\theta,t) := X(\gamma(\theta,t))$, leads to "test functions" $\phi(\theta,t)$ in formula* (2.22) *such that the functions*

$$\phi_\nu(\theta) := \frac{\partial^\nu}{\partial t^\nu}\phi(\theta,t)\big|_{t=0}, \quad \nu = 0,1,2,\ldots,n,$$

can arbitrarily be prescribed as 2π-periodic functions of class C^∞.

Proof Let us first check that, given $\phi_0,\phi_1,\ldots,\phi_n$, the computation of σ, and so of γ, can be carried out in a formal way. Consider the Fourier expansion of the function $\sigma(\theta,t)$ which is to be determined:

$$\sigma(\theta,t) = \frac{1}{2}a_0(t) + \sum_{k=1}^\infty [a_k(t)\cos k\theta + b_k(t)\sin k\theta]. \tag{2.23}$$

From $\sigma(\theta,0) = 0$ it follows that

$$a_0(0) = a_k(0) = b_k(0) = 0 \quad \text{for } k \in \mathbb{N}.$$

Furthermore,

$$\sigma_\nu(\theta) := \frac{\partial^\nu}{\partial t^\nu}\sigma(\theta,0) = \frac{1}{2}a_0^{(\nu)}(0) + \sum_{k=1}^\infty [a_k^{(\nu)}(0)\cos k\theta + b_k^{(\nu)}(0)\sin k\theta]. \tag{2.24}$$

Hence if $D_t^\nu\sigma(\theta,0)$ are known for $\nu = 1,2,\ldots,n$, one also knows all derivatives $D_\theta D_t^\nu\sigma(\theta,0) = \sigma_\nu'(\theta)$ from the defining (2.18) for σ which amounts to

$$\phi(\theta,t) = \frac{\sigma_t(\theta,t)}{1+\sigma_\theta(\theta,t)}.$$

By differentiation with respect to t we obtain

$$\phi_t = \frac{\sigma_{tt}}{1+\sigma_\theta} - \frac{\sigma_t\sigma_{\theta t}}{(1+\sigma_\theta)^2},$$

$$\phi_{tt} = \frac{\sigma_{ttt}}{1+\sigma_\theta} - \frac{2\sigma_{tt}\sigma_{\theta t}}{(1+\sigma_\theta)^2} - \frac{\sigma_t\sigma_{\theta tt}}{(1+\sigma_\theta)^2} + \frac{2\sigma_t(\sigma_{t\theta})^2}{(1+\sigma_\theta)^3}$$

etc. Setting $t = 0$ and observing that $\sigma_\theta(\theta,0) = 0$ it follows that

$$\sigma_1 = \phi_0 = \phi,$$

$$\sigma_2 = \phi_1 + \sigma_1\sigma_1',$$

$$\sigma_3 = \phi_2 + 2\sigma_2\sigma_1' + \sigma_1\sigma_2' - 2\sigma_1(\sigma_1')^2,$$

$$\dots$$

$$\sigma_{\nu+1} = \phi_\nu + f_\nu(\sigma_1, \dots, \sigma_\nu, \sigma_1', \dots, \sigma_\nu').$$

Here f_ν is a polynomial in the variables $\sigma_1, \dots, \sigma_\nu, \sigma_1', \dots, \sigma_\nu'$. This shows that, given $\phi_0, \phi_1, \dots, \phi_n$, we can successively determine $\sigma_1, \sigma_2, \dots, \sigma_{n+1}$. On account of (2.23) we then obtain

$$A_0^\nu := a_0^{(\nu)}(0), \qquad A_k^\nu := a_k^{(\nu)}(0), \qquad B_k^\nu := b_k^{(\nu)}(0) \quad \text{for } k \in \mathbb{N}.$$

Defining

$$a_k(t) := \sum_{\nu=1}^{n+1} \frac{1}{\nu!} A_k^\nu t^\nu, \qquad b_k(t) := \sum_{\nu=1}^{n+1} \frac{1}{\nu!} B_k^\nu t^\nu,$$

(2.23) furnishes the function $\gamma(\theta, t) = \theta + \sigma(\theta, t)$ with the desired properties. Furthermore, the construction shows that this procedure leads to a C^∞-function σ that is 2π-periodic with respect to θ. □

Let us inspect a variation $\hat{Z}(t) = H(Z(t))$ of a minimal surface $\hat{X} \in C^\infty(\overline{B}, \mathbb{R}^3)$ as we have just discussed. It is the harmonic extension of a variation $Z(t)$ of the boundary values X of $\hat{X}$, given by (2.15) and (2.16). Clearly, $\hat{Z}(t)$ is not merely an "inner variation" of $\hat{X}$, generated as a reparametrization $\hat{X} \circ \sigma(t)$ with a perturbation $\sigma(t) = id_{\overline{B}} + t\lambda + \cdots$ of the identity $id_{\overline{B}}$ on $\overline{B}$, but the image $\hat{Z}(t)(B)$ will differ from the image $\hat{X}(B)$. Only the images $Z(t)(S^1)$ and $X(S^1)$ of the boundary $S^1 = \partial B$ will be the same set Σ, but described by different parametrizations $Z(t) : S^1 \to \Sigma$ and $X : S^1 \to \Sigma$.

Definition 2.1 We call such a variation $\hat{Z}(t)$ a **boundary preserving variation of** $\hat{X}$ (for $|t| \ll 1$).

Note: If $\hat{X} \in \mathcal{C}(\Gamma)$ then any boundary preserving variation $\hat{Z}(t)$ (with $|t| \ll 1$) lies in $\mathcal{C}(\Gamma)$.

Definition 2.2 We say that $\hat{X}$ **is a weak relative minimizer of** D (with respect to its own boundary) if $E(0) \le E(t)$ holds for any variation $E(t) = D(\hat{Z}(t))$ of D by an arbitrary boundary preserving variation $\hat{Z}(t)$ of $\hat{X}$ with $|t| \ll 1$.

If $\hat{X} \in \mathcal{C}(\Gamma)$ is a weak relative minimizer of D in $\mathcal{C}(\Gamma)$ with respect to some C^k-norm on $\overline{B}$, then $\hat{X}$ clearly is a weak relative minimizer of D in the sense of Definition 2.2.

Let us return to formula (2.19) which states that

$$\frac{\partial}{\partial t} Z(t) = \phi(t) Z(t)_\theta.$$

According to (2.5) we have

$$Z(t)_\theta = 2\,\mathrm{Re}[iw\hat{Z}_w(w,t)]\big|_{w=e^{i\theta}},$$

and since ϕ is real-valued it follows that

$$\frac{\partial}{\partial t} Z(\theta, t) = 2\,\mathrm{Re}[iw\hat{Z}_w(w,t)\phi(\theta,t)]\big|_{w=e^{i\theta}}. \tag{2.25}$$

Since $\frac{\partial}{\partial t}$ and the harmonic extension H commute we obtain

$$\frac{\partial}{\partial t}\hat{Z}(t) = H\{2\,\mathrm{Re}[iw\hat{Z}(t)_w\phi(t)]\} \quad \text{in } \overline{B} \tag{2.26}$$

having for brevity dropped the w, except for the factor iw (as this would require a clumsy notation). Then, by

$$\frac{\partial}{\partial t}\frac{\partial}{\partial w}\hat{Z}(t) = \frac{\partial}{\partial w}\frac{\partial}{\partial t}\hat{Z}(t),$$

it follows that

$$\frac{\partial}{\partial t}\hat{Z}(t)_w = \Big(H\{2\,\mathrm{Re}[iw\hat{Z}(t)_w\phi(t)]\}\Big)_w. \tag{2.27}$$

Now a straightforward differentiation of (2.22) yields

$$\begin{aligned}\frac{d^2}{dt^2}E(t) &= 4\,\mathrm{Re}\int_{S^1} w\left\{\frac{\partial\hat{Z}(t)}{\partial t}\right\}_w \cdot \hat{Z}(t)_w\phi(t)\,dw \\ &\quad + 2\,\mathrm{Re}\int_{S^1} w\hat{Z}(t)_w\cdot\hat{Z}(t)_w\phi_t(t)\,dw.\end{aligned} \tag{2.28}$$

From (2.22) and (2.28) we obtain

Proposition 2.1 *Since $\hat{X} = \hat{Z}(0)$ is a minimal surface we have*

$$\frac{dE}{dt}(0) = 0 \tag{2.29}$$

and

$$\frac{d^2E}{dt^2}(0) = 4\,\mathrm{Re}\int_{S^1} w\left\{\frac{\partial\hat{X}}{\partial t}\right\}_w \cdot \hat{X}_w\tau\,dw \tag{2.30}$$

with $\tau := \phi(0)$. If τ is the generator of an inner forced Jacobi field attached to $\hat{X}$, then

$$\frac{d^2E}{dt^2}(0) = 0. \tag{2.31}$$

This means that

$$\delta^2 D(\hat{X}, \hat{h}) = 0 \quad \textit{for all } \hat{h} \in J_0(\hat{X}), \tag{2.32}$$

i.e. for all inner forced Jacobi fields $\hat{h} = \mathrm{Re}[iwX_w(w)\tau(w)]$.

Proof We have $\hat{X}_w \cdot \hat{X}_w = 0$ since $\hat{X}$ is a minimal surface, and so (2.29) and (2.30) are proved. Secondly, $\hat{h}$ is holomorphic in B, as it is an inner forced Jacobi field, and the w-derivative of any harmonic mapping is holomorphic whence $\{\frac{\partial\hat{X}}{\partial t}\}_w$ is holomorphic in B. Thus the integrand of $\int_{S^1}(\dots)\,dw$ in (2.30) is holomorphic. Hence this integral vanishes, since Cauchy's integral theorem implies $\int_{\partial B_r(0)}(\dots)\,dw = 0$ for any $r \in (0,1)$ and then $\int_{S^1}(\dots)\,dw = \lim_{r\to 1-0}\int_{\partial B_r(0)}(\dots)\,dw = 0$ as the integrand $(\dots)$ is continuous (and even of class C^∞) on $\overline{B}$. □

Now we want to compute $\frac{d^3}{dt^3}E(t)$, and in particular $\frac{d^3E}{dt^3}(0)$ if $\tau = \phi(0)$ is the generator of an inner forced Jacobi field. Differentiating (2.28) it follows

$$\begin{aligned}
\frac{d^3}{dt^3}E(t) &= 4\,\mathrm{Re}\int_{S^1} w\left\{\frac{\partial \hat{Z}(t)}{\partial t}\right\}_w \cdot \left\{\frac{\partial \hat{Z}(t)}{\partial t}\right\}_w \phi(t)\,dw \\
&\quad + 4\,\mathrm{Re}\int_{S^1} w\left\{\frac{\partial^2 \hat{Z}(t)}{\partial t^2}\right\}_w \cdot \hat{Z}(t)_w \phi(t)\,dw \\
&\quad + 8\,\mathrm{Re}\int_{S^1} w\left\{\frac{\partial \hat{Z}(t)}{\partial t}\right\}_w \cdot \hat{Z}(t)_w \phi_t(t)\,dw \\
&\quad + 2\,\mathrm{Re}\int_{S^1} w\hat{Z}(t)_w \cdot \hat{Z}(t)_w \phi_{tt}(t)\,dw.
\end{aligned} \tag{2.33}$$

Proposition 2.2 *Since $\hat{X} = \hat{Z}(0)$ is a minimal surface we have*

$$\frac{d^3E}{dt^3}(0) = -4\,\mathrm{Re}\int_{S^1} w^3 \hat{X}_{ww}\cdot \hat{X}_{ww}\tau^3\,dw \tag{2.34}$$

if $\tau := \phi(0)$ is the generator of an inner forced Jacobi field at $\hat{X}$.

Proof The fourth integral in (2.33) vanishes at $t = 0$ since

$$\hat{Z}(0)_w \cdot \hat{Z}(0)_w = X_w \cdot X_w = 0.$$

The integrand of the second integral in (2.33) is

$$\left\{\frac{\partial^2 \hat{Z}}{\partial t^2}(0)\right\}_w \cdot w\hat{X}_w\tau(w)$$

which is holomorphic in B since the w-derivative of a harmonic mapping is holomorphic and $\hat{h} = \mathrm{Re}[iw\hat{X}_w\tau]$ is an inner forced Jacobi field. So also the second integral in (2.33) vanishes on account of Cauchy's integral theorem. Next, using (2.27), we obtain

$$\left\{\frac{\partial}{\partial t}\hat{Z}(t)\right\}_w\bigg|_{t=0} = 2\frac{\partial}{\partial w}H\left\{\mathrm{Re}[iw\hat{X}_w\tau]\right\} = [iw\hat{X}_w\tau]_w. \tag{2.35}$$

This implies

$$\begin{aligned}
&\left[w\left\{\frac{\partial}{\partial t}\hat{Z}(t)\right\}_w \cdot \hat{Z}(t)_w\right]\bigg|_{t=0} \\
&= w[iw\hat{X}_w\tau]_w \cdot \hat{X}_w \\
&= iw\hat{X}_w \cdot \hat{X}_w\tau + iw^2\hat{X}_{ww}\cdot \hat{X}_w\tau + iw^2\hat{X}_w\cdot \hat{X}_w\tau_w = 0
\end{aligned}$$

since $\hat{X}_w \cdot \hat{X}_w = 0$, which also yields $\hat{X}_{ww} \cdot \hat{X}_w = 0$. Thus

$$\left[w \left\{ \frac{\partial}{\partial t} \hat{Z}(t) \right\}_w \cdot \hat{Z}(t)_w \right] \Big|_{t=0} = 0 \tag{2.36}$$

and so the third integral in (2.33) vanishes for $t = 0$. Finally, by (2.35),

$$\begin{aligned}
&\left(\left\{ \frac{\partial}{\partial t} \hat{Z}(t) \right\}_w \cdot \left\{ \frac{\partial}{\partial t} \hat{Z}(t) \right\}_w \right) \Big|_{t=0} \\
&= [iw\hat{X}_w\tau]_w \cdot [iwX_w\tau]_w \\
&= [i\hat{X}_w\tau + iw\hat{X}_{ww}\tau + iw\hat{X}_w\tau_w] \cdot [i\hat{X}_w\tau + iw\hat{X}_{ww}\tau + iw\hat{X}_w\tau_w] \\
&= -w^2\hat{X}_{ww} \cdot \hat{X}_{ww}\tau^2,
\end{aligned}$$

using again $\hat{X}_w \cdot \hat{X}_w = 0$ and $\hat{X}_w \cdot \hat{X}_{ww} = 0$, i.e.

$$\left(\left\{ \frac{\partial}{\partial t} \hat{Z}(t) \right\}_w \cdot \left\{ \frac{\partial}{\partial t} \hat{Z}(t) \right\}_w \right) \Big|_{t=0} = -w^2\hat{X}_{ww} \cdot \hat{X}_{ww}\tau^2. \tag{2.37}$$

Thus the first integral in (2.33) amounts to

$$-4 \operatorname{Re} \int_{S^1} w^3 \hat{X}_{ww} \cdot \hat{X}_{ww} \tau^3 dw. \qquad \square$$

In order to simplify notation we drop the t in (2.33) and write

$$\begin{aligned}
\frac{d^3}{dt^3} E = \operatorname{Re} \Big[4 \int_{S^1} w\hat{Z}_{tw} \cdot \hat{Z}_{tw}\phi \, dw + 4 \int_{S^1} w\hat{Z}_{ttw} \cdot \hat{Z}_w\phi \, dw \\
+ 8 \int_{S^1} w\hat{Z}_{tw} \cdot \hat{Z}_w\phi_t \, dw + 2 \int_{S^1} w\hat{Z}_w \cdot \hat{Z}_w\phi_{tt} \, dw \Big].
\end{aligned}$$

Differentiation yields

$$\begin{aligned}
\frac{d^4}{dt^4} E &= \operatorname{Re} \Big[12 \int_{S^1} w\hat{Z}_{ttw} \cdot \hat{Z}_{tw}\phi \, dw + 4 \int_{S^1} w\hat{Z}_{tttw} \cdot \hat{Z}_w\phi \, dw \\
&\quad + 12 \int_{S^1} w\hat{Z}_{tw} \cdot \hat{Z}_{tw}\phi_t \, dw + 12 \int_{S^1} w\hat{Z}_{ttw} \cdot \hat{Z}_w\phi_t \, dw \\
&\quad + 12 \int_{S^1} w\hat{Z}_{tw} \cdot \hat{Z}_w\phi_{tt} \, dw + 2 \int_{S^1} w\hat{Z}_w \cdot \hat{Z}_w\phi_{ttt} \, dw \Big] \\
&= \operatorname{Re}[I_1 + I_2 + I_3 + I_4 + I_5 + I_6].
\end{aligned} \tag{2.38}$$

We have $I_6(0) = 0$ since $\hat{Z}_w(0) \cdot \hat{Z}_w(0) = \hat{X}_w \cdot \hat{X}_w = 0$. Moreover, by Cauchy's theorem, $I_2(0) = 0$ since both $\hat{Z}_{tttw}\big|_{t=0} = [\hat{Z}_{ttt}(0)]_w$ and $w\hat{X}_w\tau$ are holomorphic. On account of (2.36) we also get $I_5(0) = 0$. Finally, taking (2.17) into account, we see that

$$I_3(0) = -12 \int_{S^1} w^3 \hat{X}_{ww} \cdot \hat{X}_{ww} \tau^2 \phi_t(0) \, dw,$$

and we arrive at

Proposition 2.3 *Since $\hat{X} = \hat{Z}(0)$ is a minimal surface we have*

$$\begin{aligned}\frac{d^4 E}{dt^4}(0) = 12\,\mathrm{Re}\int_{S^1} \hat{Z}_{ttw}(0)\cdot[w\hat{Z}_{tw}(0)\tau + w\hat{X}_w\phi_t(0)]\,dw \\ - 12\,\mathrm{Re}\int_{S^1} w^3\hat{X}_{ww}\cdot\hat{X}_{ww}\tau^2\phi_t(0)\,dw, \qquad (2.39)\end{aligned}$$

provided that $\tau = \phi(0)$ is the generator of an inner forced Jacobi field at $\hat{X}$.

Finally, as an exercise, we even compute $\frac{d^5 E}{dt^5}(0)$. Differentiating (2.38) it follows that

$$\frac{d^5 E}{dt^5} = \mathrm{Re}\sum_{j=1}^{9} I_j \qquad (2.40)$$

with

$$\begin{aligned}
I_1 &:= 16\int_{S^1} w\hat{Z}_{tttw}\cdot\hat{Z}_{tw}\phi\,dw, & I_2 &:= 12\int_{S^1} w\hat{Z}_{ttw}\cdot\hat{Z}_{ttw}\phi\,dw,\\
I_3 &:= 4\int_{S^1} w\hat{Z}_{ttttw}\cdot\hat{Z}_{w}\phi\,dw, & I_4 &:= 16\int_{S^1} w\hat{Z}_{tttw}\cdot\hat{Z}_{w}\phi_t\,dw,\\
I_5 &:= 48\int_{S^1} w\hat{Z}_{ttw}\cdot\hat{Z}_{tw}\phi_t\,dw, & I_6 &:= 24\int_{S^1} w\hat{Z}_{ttw}\cdot\hat{Z}_{w}\phi_{tt}\,dw,\\
I_7 &:= 24\int_{S^1} w\hat{Z}_{tw}\cdot\hat{Z}_{tw}\phi_{tt}\,dw, & I_8 &:= 16\int_{S^1} w\hat{Z}_{tw}\cdot\hat{Z}_{w}\phi_{ttt}\,dw,\\
I_9 &:= 2\int_{S^1} w\hat{Z}_{w}\cdot\hat{Z}_{w}\phi_{tttt}\,dw.
\end{aligned}$$

$I_3(0)$ vanishes by Cauchy's theorem since both $\hat{Z}_{tttt}(0)_w$ and $w\hat{X}_w\tau$ are holomorphic provided that $\tau = \phi(0)$ is the generator of a forced Jacobi field at $\hat{X}$. Furthermore, $I_8(0) = 0$ because of (2.36), and $\hat{X}_w\cdot\hat{X}_w = 0$ implies $I_9(0) = 0$. Thus we obtain by (2.37):

Proposition 2.4 *Since $\hat{X}$ is a minimal surface we have*

$$\begin{aligned}\frac{d^5 E}{dt^5}(0) = 16\,\mathrm{Re}\int_{S^1} \hat{Z}_{tttw}(0)\cdot[w\hat{Z}_{tw}(0)\tau + w\hat{X}_w\phi_t(0)]\,dw \\ + 12\,\mathrm{Re}\int_{S^1} Z_{ttw}(0)\cdot[w\hat{Z}_{ttw}(0)\tau \\ + 4w\hat{Z}_{tw}(0)\phi_t(0) + 2w\hat{X}_w\phi_{tt}(0)]\,dw \\ - 24\,\mathrm{Re}\int_{S^1} w^3\hat{X}_{ww}\cdot\hat{X}_{ww}\tau^2\phi_{tt}(0)\,dw \qquad (2.41)\end{aligned}$$

provided that $\tau = \phi(0)$ is the generator of an inner forced Jacobi field at $\hat{X}$.

Note also that in (2.39) and (2.41) we can express $\hat{Z}_{tw}(0)$ by (2.35) which we write as

$$\hat{Z}_{tw}(0) = [iw\hat{X}_w\tau]_w. \tag{2.42}$$

The values of $E''(0)$ and $E'''(0)$ in (2.30) and (2.34) depend only on $\tau = \phi(0)$ and not on any derivatives of $\phi(t)$ at $t = 0$; in this sense we say that $E''(0)$ and $E'''(0)$ are *intrinsic*. As we shall see later, this reflects important facts, namely: The Dirichlet integral D has an intrinsic second derivative d^2D, and an intrinsic third derivative d^3D in the direction of forced Jacobi fields.

Let us try to show that a nonplanar weak relative minimizer $\hat{X}$ of D cannot have a branch point in $\overline{B}$. To achieve this goal, a somewhat naive approach would be to compute sufficiently many derivatives $E^{(j)}(0) := \frac{d^jE}{dt^j}(0)$ and to hope that one can find some first non-vanishing derivative, say, $E^{(L)}(0) \neq 0$, whereas $E^{(j)}(0) = 0$ for $j = 1, 2, \ldots, L-1$. Then Taylor's formula with Cauchy's remainder term yields

$$E(t) = E(0) + \frac{1}{L!}E^{(L)}(\vartheta t)t^L \quad \text{for } |t| \ll 1,\ 0 < \vartheta < 1,$$

that is,

$$D(\hat{Z}(t)) = D(\hat{X}) + \frac{1}{L!}E^{(L)}(\vartheta t)t^L,$$

and we infer for some t with $0 < |t| \ll 1$ that

(i) $D(\hat{Z}(t)) < D(\hat{X})$ if L odd $= 2\ell + 1 \geq 3$ and $E^{(2\ell+1)}(0) \neq 0$,

and

(ii) $D(\hat{Z}(t)) < D(\hat{X})$ if L even $= 2\ell \geq 4$ and $E^{(2\ell)}(0) < 0$.

Let us see under which assumption on $\hat{X}$ this approach works for $L = 3$. Note that an arbitrary branch point $w_0 \in B$ of a minimal surface $\hat{X}$ can be moved to the origin by means of a suitable conformal automorphism of $\overline{B}$. Hence it is sufficient for our purposes to show that a minimizer $\hat{X}$ of D in $\mathcal{C}(\Gamma)$ does not have $w = 0$ as a branch point. Therefore we shall from now on assume the following **normal form of a nonplanar minimal surface** $\hat{X}$ (cf. Dierkes, Hildebrandt and Sauvigny [1], Sect. 3.2):

$\hat{X}$ has $w = 0$ as a branch point of *order* n, i.e.

$$\hat{X}_w(w) = aw^n + o(w^n) \quad \text{as } w \to 0.$$

Choosing a suitable Cartesian coordinate system in $\mathbb{R}^3$ we may assume that $\hat{X}_w$ can be written as

$$\hat{X}_w(w) = (A_1w^n + A_2w^{n+1} + \cdots, R_mw^m + R_{m+1}w^{m+1} + \cdots), \quad m > n, \tag{2.43}$$

with $A_j \in \mathbb{C}^2$, $R_j \in \mathbb{C}$, $A_1 \neq 0$ and $R_m \neq 0$ for some integer m satisfying $m > n$; the number m is called the *index* of the branch point $w = 0$ of $\hat{X}$ given in the normal form (2.43). Note that a surface $\hat{X}$ can also be brought into the normal form (2.43) (with $n = 0$) if $\hat{X}$ is regular at $w = 0$.

Lemma 2.2 *The normal form* (2.43) *satisfies*

$$\begin{aligned} &A_1 \cdot A_1 = 0, \qquad A_k = \lambda_k \cdot A_1 \quad \text{for } k = 1, 2, \ldots, 2(m-n), \\ &A_1 \cdot A_{2m-2n+1} = -\frac{1}{2} R_m^2, \end{aligned} \tag{2.44}$$

and therefore

$$\hat{X}_{ww}(w) \cdot \hat{X}_{ww}(w) = (m-n)^2 R_m^2 w^{2m-2} + \cdots, \qquad R_m \neq 0. \tag{2.45}$$

Proof Equation (2.43) implies

$$\hat{X}_w(w) \cdot \hat{X}_w(w) = (w^{2n} p(w) + R_m^2 w^{2m}) + O(|w|^{2m+1}) \quad \text{as } w \to 0$$

where $p(w)$ is a polynomial of degree 2ℓ in w with $\ell := m - n$ which is of the form

$$\begin{aligned} p(w) = {} & A_1 \cdot A_1 + 2A_1 \cdot A_2 w + (2A_1 \cdot A_3 + A_2 \cdot A_2) w^2 \\ & + (2A_1 \cdot A_4 + 2A_2 \cdot A_3) w^3 + (2A_1 \cdot A_5 + 2A_2 \cdot A_4 + A_3 \cdot A_3) w^4 \\ & + \cdots + (2A_1 \cdot A_{2\ell+1} + 2A_2 \cdot A_{2\ell} + \cdots + 2A_{\ell+2} \cdot A_\ell + A_{\ell+1} \cdot A_{\ell+1}) w^{2\ell} \\ = {} & c_0 + c_1 w + c_2 w^2 + \cdots + c_{2\ell} w^{2\ell}, \quad c_j \in \mathbb{C}. \end{aligned}$$

Since $\hat{X}_w \cdot \hat{X}_w = 0$ we obtain

$$c_0 = c_1 = \cdots = c_{2\ell-1} = 0, \quad c_{2\ell} + R_m^2 = 0.$$

Let $\langle A', A'' \rangle := A' \cdot \overline{A''}$ be the Hermitian scalar product of two vectors $A', A'' \in \mathbb{C}^2$. The two equations $c_0 = 0$ and $c_1 = 0$ yield $A_1 \cdot A_1 = 0$ and $A_1 \cdot A_2 = 0$ which are equivalent to

$$\langle A_1, \overline{A}_1 \rangle = 0 \quad \text{and} \quad \langle A_2, \overline{A}_1 \rangle = 0.$$

Since $A_1 \neq 0$ and $\overline{A}_1 \neq 0$ this implies

$$A_2 = \lambda_2 A_1 \quad \text{for some } \lambda_2 \in \mathbb{C},$$

and so we also obtain

$$A_2 \cdot A_2 = \lambda_2^2 A_1 \cdot A_1 = 0.$$

On account of $c_2 = 0$ it follows $A_1 \cdot A_3 = 0$, and thus it follows

$$\langle A_1, \overline{A}_1 \rangle = 0 \quad \text{and} \quad \langle A_3, \overline{A}_1 \rangle = 0$$

whence

$$A_3 = \lambda_3 A_1 \quad \text{for some } \lambda_3 \in \mathbb{C},$$

and so

$$A_2 \cdot A_3 = \lambda_2 \lambda_3 A_1 \cdot A_1 = 0.$$

Then $c_3 = 0$ yields $A_1 \cdot A_4 = 0$, therefore

$$\langle A_1, \overline{A}_1 \rangle = 0 \quad \text{and} \quad \langle A_4, \overline{A}_1 \rangle = 0;$$

consequently

$$A_4 = \lambda_4 A_1 \quad \text{for some } \lambda_4 \in \mathbb{C}.$$

In this way we proceed inductively using $c_0 = 0, \ldots, c_{2\ell-1} = 0$ and obtain $A_k = \lambda_k A_1$ for $k = 1, 2, \ldots, 2(m-n)$. Since $A_1 \cdot A_1 = 0$ it follows that

$$A_j \cdot A_k = 0 \quad \text{for } 1 \le j,\ k \le 2(m-n). \tag{2.46}$$

Then the equation $c_{2\ell} + R_m^2 = 0$ implies $2A_1 \cdot A_{2\ell+1} + R_m^2 = 0$, i.e.

$$A_1 \cdot A_{2(m-n)+1} = -\frac{1}{2} R_m^2. \tag{2.47}$$

Furthermore, from

$$\hat{X}_w(w) = (A_1 w^n + A_2 w^{n+1} + \cdots + A_{2m-2n+1} w^{2m-n} + \cdots,\ R_m w^m + \cdots)$$

we infer

$$\begin{aligned}\hat{X}_{ww}(w) = (&nA_1 w^{n-1} + \cdots + (2m-n)A_{2m-2n+1} w^{2m-n-1} + \cdots,\\ &mR_m w^{m-1} + \cdots).\end{aligned}$$

Then (2.46) implies

$$\hat{X}_{ww}(w) \cdot \hat{X}_{ww}(w) = [2n(2m-n)A_1 \cdot A_{2m-2n+1} + m^2 R_m^2] w^{2m-2} + \cdots,$$

and by (2.47) we arrive at

$$\hat{X}_{ww}(w) \cdot \hat{X}_{ww}(w) = [-n(2m-n)R_m^2 + m^2 R_m^2] w^{2m-2} + \cdots,$$

which is equivalent to (2.45). □

Theorem 2.1 (D. Wienholtz) *Let $\hat{X}$ be a minimal surface in normal form with a branch point at $w = 0$ which is of order n and index m, $n < m$, and suppose that $2m - 2 < 3n$ (or, equivalently, $2m + 2 \le 3(n+1)$). Then we can choose a generator τ of a forced Jacobi field $\hat{h}$ such that $E^{(3)}(0) < 0$, and so $\hat{X}$ is not a weak relative minimizer of D.*

Proof Define the integer k by

$$k := (2m+2) - 2(n+1).$$

Because of $m > n$ and $2m - 2 < 3n$ it follows that

$$1 < k \le n + 1.$$

Let

$$\tau_0 := cw^{-n-1} + \bar{c}w^{n+1}, \qquad \tau_1 := cw^{-k} + \bar{c}w^k, \qquad c \in \mathbb{C},$$

and set

(i) $\tau := \tau_0$ if $k = n + 1$;
(ii) $\tau := \epsilon\tau_0 + \tau_1$, $\epsilon > 0$, if $k < n + 1$;

In both cases τ is a generator of a forced Jacobi field at $\hat{X}$, since $w\hat{X}_w(w)$ has a zero of order $n+1$ at $w=0$, and $\operatorname{Im}\tau = 0$ on ∂B. By (2.45) it follows for $w \in B$ that

$$w^3\hat{X}_{ww}(w)\cdot\hat{X}_{ww}(w) = (m-n)^2 R_m^2 w^{2m+1} + \cdots$$

where $+\cdots$ always stands for higher order terms of a convergent power series. In case (i) one has

$$\tau^3(w) = c^3 w^{-3(n+1)} + \cdots,$$

and so

$$w^3\hat{X}_{ww}(w)\cdot\hat{X}_{ww}(w)\tau(w)^3 = (m-n)^2 R_m^2 c^3 w^{-1} + f(w)$$

where $f(w)$ is holomorphic in B and continuous on $\overline{B}$. Then formula (2.34) of Proposition 2.3 in conjunction with Cauchy's integral theorem yields

$$E^{(3)}(0) = -4\operatorname{Re}[2\pi i(m-n)^2 R_m^2 c^3] \quad \text{if } k = n+1.$$

With a suitable choice of $c \in \mathbb{C}$ we can arrange for $E^{(3)}(0) < 0$ since $R_m \neq 0$ and $(m-n)^2 \geq 1$.

In case (ii) we write $w^3\hat{X}_{ww}\cdot\hat{X}_{ww}$ as

$$w^3\hat{X}_{ww}(w)\cdot\hat{X}_{ww}(w) = (m-n)^2 R_m^2 w^{2m+1} + f(w),$$

where

$$f(w) := w^{2m+2}\sum_{j=0}^{\infty} a_j w^j, \quad a_j \in \mathbb{C}.$$

From

$$\tau^3 = \epsilon^3\tau_0^3 + 3\epsilon^2\tau_0^2\tau_1 + 3\epsilon\tau_0\tau_1^2 + \tau_1^3$$

it follows that

$$g(w) := w^3\hat{X}_{ww}(w)\cdot\hat{X}_{ww}(w)\tau^3(w)$$

is meromorphic in B, continuous in $\{w : \rho < |w| \leq 1\}$ for some $\rho \in (0,1)$, and its Laurent expansion at $w=0$ has the residue

$$\operatorname{Res}_{w=0}(g) = 3\epsilon^2 c^3(m-n)^2 R_m^2 + \epsilon^3 c^3 a_{n-k}, \quad 1 < k \leq n.$$

Cauchy's residue theorem together with formula (2.34) of Proposition 2.3 then imply

$$E^{(3)}(0) = -4\operatorname{Re}\{2\pi i[3\epsilon^2 c^3(m-n)^2 R_m^2 + \epsilon^3 c^3 a_{n-k}]\} \quad \text{for } k < n+1.$$

By an appropriate choice of $c \in \mathbb{C}$ and ϵ with $0 < \epsilon < 1$ we can achieve that $E^{(3)}(0) < 0$ also in case (ii). □

The following definition will prove to be very useful.

Definition 2.3 Let $\hat{X}$ be a minimal surface in normal form having $w=0$ as a branch point of order n and of index m. Then $w=0$ is called an **exceptional branch point** if $m+1 = \kappa(n+1)$ for some $\kappa \in \mathbb{N}$; necessarily $\kappa > 1$.

Remark 2.1 If $2m-2<3n$, i.e. $2(m+1)\le 3(n+1)$, then $w=0$ is not exceptional, because $(m+1)=\kappa(n+1)$ with $\kappa>1$ implies $2\kappa(n+1)\le 3(n+1)$ and therefore $2\kappa\le 3$ which is impossible for $\kappa\in\mathbb{N}$ with $\kappa>1$.

Remark 2.2 Now we want to show that the notion "$w=0$ *is an exceptional branch point*" is closely related to the notion "$w=0$ *is a false branch point*". To this end we choose an arbitrary minimal surface $\hat{Z}(\zeta)$, $\zeta\in B$, in normal form without $\zeta=0$ being a branch point, i.e. $\hat{Z}=\operatorname{Re} g$ where $g:B\to\mathbb{C}^3$ is holomorphic and of the form

$$g(\zeta)=\hat{Z}(0)+(B_0\zeta+B_1\zeta^2+\cdots, C_\kappa\zeta^\kappa+\cdots),\qquad B_0\neq 0,\ C_\kappa\neq 0,\ \kappa>1.$$

Consider a conformal mapping $w\mapsto\zeta=\varphi(w)$ from B into B with $\varphi(0)=0$ which is provided by a holomorphic function

$$\varphi(w)=aw+\cdots,\qquad a\neq 0,\ w\in B.$$

Then $\hat{X}(w):=\operatorname{Re} f(w)$ with $f(w):=g(\varphi^{n+1}(w))$, $w\in B$, is a minimal surface $\hat{X}:B\to\mathbb{R}^3$ such that $\hat{X}(0)=\hat{Z}(0)$ and

$$f(w)=\hat{X}(0)+(a^{n+1}B_0w^{n+1}+\cdots, a^{\kappa(n+1)}C_\kappa w^{\kappa(n+1)}+\cdots).$$

Thus we obtain for $\hat{X}_w=\frac{1}{2}f'$ that

$$\hat{X}_w(w)=(A_1w^n+\cdots, R_m w^m+\cdots),\qquad A_1\neq 0,\ R_m\neq 0,$$

and so $\hat{X}(w)$, $w\in B$, is a minimal surface in normal form which has the branch point $w=0$ of order n and index $m:=\kappa(n+1)-1$, whence $w=0$ is *exceptional*. Clearly $\hat{X}$ is obtained from the minimal immersion $\hat{Z}(\zeta)$ as a *false branch point* by setting $\hat{X}:=\hat{Z}\circ\varphi^{n+1}$. As the "false parametrization" $\hat{X}$ of the regular surface $\mathcal{S}:=\hat{Z}(B)$ is produced by an analytic expression $\zeta=\varphi^{n+1}(w)$ we call $w=0$ an "*analytic false branch point*".

Let $\hat{X}$ be a minimal surface with $w=0$ a branch point of order n. Now, if we know that the image under $\hat{X}$ of a small neighbourhood U of 0 is an analytic regular (embedded) surface $\mathcal{S}$, then $w=0$ is a false branch point and it is not hard to see that $w=0$ is also analytically false.

To this end, let $Y:U\to\mathcal{S}$ be a $C^{2,\alpha}$ smooth regular conformal parametrization of f. Then

$$\varphi:=Y^{-1}\circ X$$

is conformal and we may presume holomorphic. Since $\hat{X}$ has a branch point of order n, φ locally has the form

$$\varphi(z)=a_{n+1}w^{n+1}+\cdots,$$

where $a_{n+1}\neq 0$.

Therefore, there is a holomorphic function ψ defined on a neighbourhood $V\subset U$ of 0 such that $\varphi=\psi^{n+1}$ and we may assume $\psi:V\to\psi(V)$ is biholomorphic. Then

$$Y^{-1}\circ\hat{X}=\psi^{n+1}$$

implying that

$$Y^{-1} \circ \hat{X} \circ \psi^{-1}(w) = w^{n+1}$$

or

$$\hat{X} \circ \psi^{-1}(w) = Y(w^{n+1})$$

i.e. $w = 0$ is locally analytically false.

In Remark 2.1 we have noted that $w = 0$ cannot be "exceptional" if $2m - n < 3n$, and so it cannot be an "analytic false branch point".

It will be useful to have a **characterization of the non-exceptional branch points**, the proof of which is left to the reader.

Lemma 2.3 *The branch point $w = 0$ is non-exceptional if and only if one of the following two conditions is satisfied*:

(i) *There is an even integer L with*

$$(L-1)(n+1) < 2(m+1) < L(n+1). \tag{2.48}$$

(ii) *There is an odd integer L with*

$$(L-1)(n+1) < 2(m+1) \le L(n+1). \tag{2.49}$$

We say that $w = 0$ satisfies condition (T_L) if either (2.48) *with L even or* (2.49) *with L odd holds.*

In Theorem 2.1 it was shown that $E^{(3)}(0)$ can be made negative if $2m - 2 < 3n$. Therefore *we shall now assume that* $2m - 2 \ge 3n$. It takes some experience to realize that the right approach to success lies in separating the two cases "$w = 0$ *is non-exceptional*" and "$w = 0$ *is exceptional*". Instead one might guess that the right generalization of Wienholtz's theorem consists in considering the cases

$$(L-1)n \le 2m - 2 < Ln, \quad L \in \mathbb{N}, \text{ with } L \ge 3 \tag{C_L}$$

and hoping that one can prove

$$E^{(j)}(0) = 0 \quad \text{for } 1 \le j \le L-1, \quad E^{(L)}(0) < 0$$

using appropriate choices of forced Jacobi fields in varying the minimal surface $\hat{X}$. Unfortunately this is not the case. To see what happens we study the two cases

$$3n \le 2m - 2 < 4n \tag{C_4}$$

and

$$4n \le 2m - 2 < 5n \tag{C_5}$$

by computing $E^{(4)}(0)$ in the first case and $E^{(5)}(0)$ in the second one. We begin by treating special cases of (C_4) and (C_5), where we can proceed in a similar way as before with $E^{(3)}(0)$ for $2n \le 2m - 2 < 3n$.

The case (C_4) **with** $2m - 2 = 4p$, $p \in \mathbb{N}$.

Proposition 2.5 *If $w\hat{Z}_{tw}(0)\tau + w\hat{X}_w\phi_t(0)$ is holomorphic, then*

$$E^{(4)}(0) = -12\,\mathrm{Re}\int_{S^1} w^3\hat{X}_{ww}\cdot\hat{X}_{ww}\tau^2\phi_t(0)\,dw. \tag{2.50}$$

Proof Since $\hat{Z}_{ttw}(0)$ is holomorphic in B, the integrand of the first integral in (2.39) is holomorphic, and so this integral vanishes. □

Remark 2.3 In case (C_4) with $2m-2=4p$ the branch point $w=0$ is non-exceptional. To see this we note that $p<n$ whence

$$2m+2=4(p+1)<4(n+1)$$

and therefore

$$n+1<m+1<2(n+1).$$

Also note that $n=1,2,3$ are not possible since $n=1$ would imply $p<1$; $n=2$ would mean $p=1$ whence $6=3n\le 4p=4$; and $n=3$ would imply $p\le 2$, and so $9=3n\le 4p=8$. Finally $3n\le 4p$ and $n\ge 4$ yields $p\ge 3$.

Theorem 2.2 *If $3n\le 2m-2=4p<4n$ for some $p\in\mathbb{N}$, then one can find a variation $\hat{Z}(t)$ of $\hat{X}$ such that $E^{(4)}(0)<0$, whereas $E^{(j)}(0)=0$ for $j=1,2,3$.*

Proof First we want to choose $\tau=\phi(0)$ and $\phi_t(0)$ in such a way that the assumption of Proposition 2.5 is satisfied. To this end, set

$$\tau(w):=(a-ib)w^{-p-1}+(a+ib)w^{p+1},$$

which clearly is a generator of a forced Jacobi field. By (2.43) we get

$$\begin{aligned}
&w\hat{X}_w(w)\tau(w)\\
&\quad=(a-ib)(A_1w^{n-p}+A_2w^{n-p+1}+\cdots+A_{2m-2n+1}w^{2m-n-p}+\cdots,\\
&\qquad R_mw^{m-p}+\cdots)+(a+ib)(A_1w^{n+p+2}+\cdots,R_mw^{m+p+2}+\cdots).
\end{aligned}$$

By (2.35) it follows

$$\begin{aligned}
&w\hat{Z}_{tw}(w,0)\tau(w)\\
&\quad=w[iw\hat{X}_w(w)\tau(w)]_w\tau(w)\\
&\quad=i(a-ib)^2((n-p)A_1w^{n-2p-1}+(n-p+1)A_2w^{n-2p}+\cdots\\
&\qquad+(2m-n-p)A_{2m-2n+1}w^{2m-n-2p-1}+\cdots,(m-p)R_mw^{m-2p-1}+\cdots).
\end{aligned}$$

Note that $2m-2=4p$ implies $m-2p-1=0$, whence $n-2p-1<0$ because of $m>n$, but $2m-n-2p-1=(m-2p-1)+(m-n)=m-n>0$. Thus the third component above has no pole, while the first (vectorial) component has a pole at least in the first term, but no pole anymore from the $(2m-2n+1)^{\text{th}}$ term on. These poles will be removed by adding $w\hat{X}_w\phi_t(0)$ to $w\hat{Z}_{tw}(0)\tau$ with an appropriately chosen value of $\phi_t(0)$. We set

$$\phi_t(0) := \sum_{\ell=1}^{s} \psi_\ell$$

with

$$\begin{aligned}\psi_1(w) :=& -i(n-p)(a-ib)^2\lambda_\ell w^{-2p-2}\\ &+ i(n-p)(a+ib)^2\overline{\lambda}_\ell w^{2p+2}\end{aligned}$$

where $A_k = \lambda_k A_1$ (cf. Lemma 2.2). The number s is the index of the last term $(n-p+s)A_{s+1}w^{n-2p+s-1}$ where $n-2p+s-1$ is non-negative. Now $w\hat{Z}_{tw}(0)\tau + w\hat{X}_w\psi_1$ has no pole associated to A, and poles of the same order or less associated to A_k, $k \le s$. Choose ψ_2, so that there is no pole associated to A_2. Continue to define ψ_ℓ so that all poles are removed.

Note that

$$w\hat{X}_w(w) = (A_1 w^{n+1} + A_2 w^{n+2} + \cdots + A_{2m-2n+1} w^{2m-n+1} + \cdots, R_m w^{m+1} + \cdots)$$

and

$$A_1 \cdot A_k = 0 \quad \text{for } k = 1, 2, \ldots, 2m-2n.$$

Therefore, $w\hat{X}_w\phi_t(0) = w\hat{X}_w \cdot [\psi_1 + \psi_2 + \cdots + \psi_s]$ removes all poles from $w\hat{Z}_{tw}(0)\tau$. Consequently $w\hat{Z}_{tw}(0)\tau + w\hat{X}_w\phi_t(0)$ is holomorphic, and so we have

$$E^{(4)}(0) = -12\,\mathrm{Re}\int_{S^1} w^3 \hat{X}_{ww} \cdot \hat{X}_{ww}\tau^2\phi_t(0)\,dw.$$

Formula (2.45) yields

$$w^3\hat{X}_{ww}(w) \cdot \hat{X}_{ww}(w) = (m-n)^2 R_m^2 w^{2m+1} + \cdots.$$

The leading term in $\phi_t(0)$ is that of ψ_1, and

$$\psi_1(w) = -i(n-p)(a-ib)^2 w^{-2p-2} + \cdots.$$

Furthermore,

$$\tau^2(w) = (a-ib)^2 w^{-2p-2} + \cdots,$$

and so

$$\tau^2(w)\phi_t(w,0) = -i(a-ib)^4(n-p)w^{-4p-4} + \cdots.$$

Noticing that $2m+1 = (2m+2)-1 = 4(p+1)-1$, and setting

$$\kappa := 12(m-n)^2(n-p) > 0$$

we obtain

$$E^{(4)}(0) = \kappa\,\mathrm{Re}\left[i(a-ib)^4 R_m^2 \int_{S^1} \frac{dw}{w}\right] = -2\pi\kappa\,\mathrm{Re}[(a-ib)^4 R_m^2]$$

and an appropriate choice of a and b yields $E^{(4)}(0) < 0$. Finally we note that $E^{(2)}(0) = 0$ and $E^{(3)}(0) = 0$ for the above choice of $\hat{Z}(t)$. The first statement follows from Proposition 2.1. To verify the second, we recall formula (2.34) from Proposition 2.2:

$$E^{(3)}(0) = -4\,\mathrm{Re} \int_{S^1} w^3 \hat{X}_{ww} \cdot \hat{X}_{ww} \tau^3 dw.$$

From the preceding computations it follows that

$$w^3 \hat{X}_{ww}(w) \cdot \hat{X}_{ww}(w) \tau^3(w) = (m-n)^2 R_m^2 (a-ib)^3 w^{2m+1-3(p+1)} + \cdots,$$

and, by assumption, $2m-2 = 4p$, whence

$$2m+1-3(p+1) = 4p+3-3(p+1) = p > 1;$$

therefore $E^{(3)}(0) = 0$. □

Remark 2.4 Under the special assumption that $2m-2 = 4p$ we were able to carry out the program outlined above for $L = 4$. However, applying the method from Theorem 2.2 to cases when $2m-2 \not\equiv 0 \bmod 4$ one seems to get nowhere. However, trying another approach similar to that used in the proof of Theorem 2.1, one is able to handle the case (C_4) under the additional assumption $2m-2 \equiv 2 \bmod 4$ by considering the next higher derivative, namely $E^{(5)}(0)$ instead of $E^{(4)}(0)$, cf. Theorem 2.4 stated later on. This seems to shatter the hope that one can always make $E^{(L)}(0)$ negative, with $E^{(j)}(0) = 0$ for $1 \le j \le L-1$, if (C_L) is satisfied. In fact, by studying assumption (C_5) we shall realize that (C_L) is probably not the appropriate classification for developing methods that in general lead to our goal. Rather the case (C_5) will teach us that one should distinguish between the cases "exceptional" and "non-exceptional" using the classification given in Lemma 2.3 for this purpose.

Let us mention that, assuming (C_4), the branch point $w = 0$ is non-exceptional according to Lemma 2.3, since $3n \le 2m-2 < 4n$ implies

$$3(n+1) < 3n+4 \le 2m+2 < 4(n+1).$$

Let us now turn to the *investigation of* (C_5) *by means of the fifth derivative* $E^{(5)}(0)$.

Lemma 2.4 *If* $f(w) := w\hat{Z}_{tw}(0)\tau + w\hat{X}_w \phi_t(0)$ *is holomorphic, then*

$$\begin{aligned} \hat{Z}_{ttw}(0) &= \{iw[iw\hat{X}_w \tau]_w \tau + iw\hat{X}_w \phi_t(0)\}_w, \\ \hat{Z}_{ttw}(0) \cdot \hat{X}_w &= -\hat{Z}_{tw}(0) \cdot \hat{Z}_{tw}(0) = w^2 \hat{X}_{ww} \cdot \hat{X}_{ww} \tau^2. \end{aligned} \tag{2.51}$$

Proof By (2.27) we have

$$\hat{Z}_{tw} = \{2H[\mathrm{Re}(iw\hat{Z}_w \phi)]\}_w$$

whence

$$\hat{Z}_{ttw} = \{2H[\mathrm{Re}(iw\hat{Z}_{tw}\phi + iw\hat{Z}_w \phi_t)]\}_w$$

and therefore

$$\begin{aligned}\hat{Z}_{ttw}(0) &= \{2H[\mathrm{Re}(if)]\}_w = \{if\}_w \\ &= \{iw\hat{Z}_{tw}(0)\tau + iw\hat{X}_w\phi_t(0)\}_w.\end{aligned}$$

By (2.35),

$$\hat{Z}_{tw}(0) = [iw\hat{X}_w\tau]_w,$$

and so

$$\hat{Z}_{ttw}(0) = \{iw[iw\hat{X}_w\tau]_w\tau + iw\hat{X}_w\phi_t(0)\}_w.$$

It follows that

$$Z_{ttw}(0)\cdot\hat{X}_w = \{iw[i\hat{X}_w\tau + iw\hat{X}_{ww}\tau + iw\hat{X}_w\tau_w]\tau + iw\hat{X}_w\phi_t(0)\}_w\cdot\hat{X}_w.$$

From $\hat{X}_w\cdot\hat{X}_w = 0$ one obtains $\hat{X}_w\cdot\hat{X}_{ww} = 0$, and then

$$\hat{X}_{www}\cdot\hat{X}_w = -\hat{X}_{ww}\cdot\hat{X}_{ww}.$$

This leads to

$$\begin{aligned}\hat{Z}_{ttw}(0)\cdot\hat{X}_w &= -w^2\hat{X}_{www}\cdot\hat{X}_w\tau^2 \\ &= w^2\hat{X}_{ww}\cdot\hat{X}_{ww}\tau^2 = -\hat{Z}_{tw}(0)\cdot\hat{Z}_{tw}(0),\end{aligned}$$

taking (2.37) into account. □

Proposition 2.4 and Lemma 2.4 imply

Proposition 2.6 *If* $f(w) := w\hat{Z}_{tw}(0)\tau + w\hat{X}_w\phi_t(0)$ *is holomorphic, then*

$$E^{(5)}(0) = 12\,\mathrm{Re}\int_{S^1}[w\hat{Z}_{ttw}(0)\cdot\hat{Z}_{ttw}(0)\tau + 4w\hat{Z}_{ttw}(0)\cdot\hat{Z}_{tw}(0)\phi_t(0)]\,dw. \quad (2.52)$$

We are now going to discuss the envisioned program for the case (C_5) using the simplified form (2.52) for the fifth derivative $E^{(5)}(0)$. It will be useful to distinguish several subcases of (C_5):

(a) $5n \le 2m+2$,
(b) $5n > 2m+2$.

In case (a) we have $5n \le 2m+2 < 5n+4$, that is,

$$2m+2 = 5n+\alpha, \quad 0\le\alpha\le 3.$$

Therefore (a) consists of the four subcases

$$2m-5n = 0,\,1,\,-1,\,-2. \quad (2.53)$$

In case (b) we have $5n > 2m+2$, and (C_5) implies $2m+2\ge n+4$, whence $5n > n+4$, and so we have $n>1$ in case (b).

Case (a) allows an easy treatment based on the following representation of $2m+2$ which we apply successively for $\alpha = 0, 1, 2, 3$ to deal with the four cases (2.53). We write

$$\alpha(n+1) + \beta n = 2m + 2$$

with $\alpha := 2m + 2 - 5n$, $\beta := 5 - \alpha$ where $0 \le \alpha \le 3$ and $\beta \ge 2$. Then we choose

$$\tau := \tau_0 + \epsilon \tau_1, \quad \epsilon > 0,$$

where

$$\tau_0 := cw^{-n} + \overline{c}w^n, \quad \tau_1 := cw^{-n-1} + \overline{c}w^{n+1}, \quad c \in \mathbb{C}.$$

With an appropriate choice of $\phi_t(0)$ we obtain by an elimination procedure similar to the one used in the proof of Theorem 2.2 that $f := w\hat{Z}_{tw}(0)\tau + w\hat{X}_w\phi_t(0)$ is holomorphic. Here and in the sequel we omit the lengthy computations and merely state the results. As f is holomorphic one can use formula (2.52) for $E^{(5)}(0)$; we investigate the four different cases of (2.53) separately, but note that always

$$E^{(j)}(0) = 0, \quad j = 1, \ldots, 4.$$

(I) $2m - 5n = 0$, $1 \le n \le 4$. Only (i) $n = 2$ and (ii) $n = 4$ are possible. This leads to

(i) $n = 2$, $m = 5$, $(m+1) = 2(n+1)$, i.e. $w = 0$ is exceptional;
(ii) $n = 4$, $m = 10$, hence $m + 1 \not\equiv 0 \bmod (n+1)$, and so $w = 0$ is not exceptional.

For (i) we obtain $E^{(5)}(0) = 0 + o(\epsilon)$, whereas (ii) yields

$$E^{(5)}(0) = 12\,\mathrm{Re}[2\pi i \cdot 360 \cdot \epsilon^2 \cdot c^5 R_m^2] + o(\epsilon^2)$$

which can be made negative by appropriate choice of c. Thus the method is inconclusive for (i), but gives the desired result for (ii).

(II) $2m - 5n = 1$, $1 \le n \le 4$. Then necessarily either (i) $n = 1$ or (ii) $n = 3$. Here,

(i) $n = 1$, $m = 3$, $m + 1 = 2(n+1)$, i.e. $w = 0$ is exceptional;
(ii) $n = 3$, $m = 8$, and $m + 1 \not\equiv 0 \bmod (n+1)$, hence $w = 0$ is not exceptional.

For (i) it follows that $E^{(5)}(0) = 0 + o(\epsilon^3)$, i.e. the method is inconclusive, while for (ii) one gets

$$E^{(5)}(0) = 12 \cdot \mathrm{Re}[2\pi i \cdot 250 \cdot \epsilon^3 \cdot c^5 R_m^2] + o(\epsilon^3),$$

and so $E^{(5)}(0) < 0$ for a suitable choice of c.

(III) $2m - 5n = -1$, $1 \le n \le 4$. Then either (i) $n = 1$ or (ii) $n = 3$, i.e.

(i) $n = 1$, $m = 2$, and so $m + 1 \not\equiv 0 \bmod (n+1)$, i.e. $w = 0$ is not exceptional;
(ii) $n = 3$, $m = 7$, whence $m + 1 = 2(n+1)$, i.e. $w = 0$ is exceptional.

For (i) we have $2m-2<3n$, and this case was already dealt with in the positive sense by using $E^{(3)}(0)$, cf. Theorem 2.1. For (ii) the method is again inconclusive since one obtains

$$E^{(5)}(0)=0+o(\epsilon).$$

(IV) $2m-5n=-2$, $1\le n\le 4$. Then either (i) $n=2$ or (ii) $n=4$, that is,

(i) $n=2$, $m=4$, whence $m+1\not\equiv 0 \bmod (n+1)$, i.e. $w=0$ is not exceptional;
(ii) $n=4$, $m=9$, and so $m+1=2(n+1)$, i.e. $w=0$ is exceptional.

In case (i) we have $3n=2m-2<4n$, i.e. condition (C_4) holds, and this case will be tackled by Theorem 2.4, to be stated later on. Case (ii) leads to $E^{(5)}(0)=0+o(1)$ as $\epsilon\to 0$ which is once again inconclusive.

Conclusion *The method is inconclusive in all of the exceptional cases. In the non-exceptional cases it either leads to the positive result $E^{(5)}(0)<0$ for appropriate choice of c, or one can apply the cases (C_3) or (C_4), and here one obtains the desired results $E^{(3)}(0)<0$ or $E^{(4)}(0)<0$ respectively (see Theorems 2.1 and 2.4).*

Now we turn to case (b). We first note that (C_5) together with (b) implies $4(n+1)\le 2m+2<5n$. Hence either (i) $2(n+1)=m+1$, or (ii) $4(n+1)<2m+2<5n$. Therefore, $w=0$ is exceptional in case (i) and non-exceptional in case (ii). Furthermore we have

$$2m+2=4n+k \quad \text{with } 4\le k<n,$$

where $k=4$ is case (i) and $4<k<n$ is case (ii).

In order to treat case (b) which in some sense is the "*general subcase*" of (C_5) we use

$$\tau:=c\cdot(\epsilon w^{-n}+w^{-k})+\overline{c}\cdot(\epsilon w^{n}+w^{k}).$$

Choosing $\phi_t(0)$ appropriately we achieve that f is holomorphic, and so $E^{(5)}(0)$ is given by (2.52). Moreover, $E^{(j)}(0)=0$ for $1\le j\le 4$. It turns out that

$$E^{(5)}(0)=12\cdot \mathrm{Re}[2\pi i c^3\epsilon^4\gamma R_m^2]+o(\epsilon^4), \quad \epsilon>0,$$

with

$$\gamma=(m-n)(k-4)^2\left[\frac{5}{4}n+\frac{5}{8}(k-2)\right]$$

and $\gamma=0$ in case (i), whereas $\gamma>0$ in case (ii).

Thus the following result is established:

Theorem 2.3 *Suppose that (C_5) and (b) hold, hence $4n+4\le 2m+2<5n$. This implies $2m+2=4n+k$ with $4\le k<n$. For $k=4$ the branch point $w=0$ is exceptional, and the method is nonconclusive. If, however, $4<k<n$, then $\tau=\phi(0)$ and $\phi_t(0)$ can be chosen in such a way that $E^{(5)}(0)<0$ and $E^{(j)}(0)=0$ for $j=1,\dots,4$.*

Next, we want to prove that the remaining cases of (C_4) lead to a conclusive result also for the remaining possibility $2m-2\neq 4p$ for some $p\in\mathbb{N}$ with $1\le p<n$. Because of $3n\le 2m-2<4n$ we can write $2m-2=4p+k$ with $0<k<4$ (the case $k = 0$ was treated before). Since k must be even, we are left with $k = 2$, and we recall that $w = 0$ is a non-exceptional branch point in case (C_4).

Theorem 2.4 *Suppose that* $3n \le 2m - 2 = 4p + 2 < 4n$ *with* $1 \le p < n$ *holds (this is the subcase of* (C_4) *that was not treated in Theorem 2.2). Then* $\tau = \phi(0)$ *and* $\phi_t(0)$ *can be chosen in such a way that*

$$E^{(j)}(0) = 0 \quad \text{for } j = 1, \ldots, 4, \qquad E^{(5)}(0) < 0.$$

Proof This follows with

$$\tau := c(w^{-k} + \epsilon w^{-p-1}) + \overline{c} \cdot (w^k + \epsilon w^{p+1}), \quad \epsilon > 0,$$

and setting

$$-\phi_t(0) := \epsilon^2 c^2 (n - p) w^{-2(p+1)} + \epsilon c^2 (2n + 1 - p - k) w^{-(p+1+k)} + \cdots.$$

Then $E^{(j)}(0) = 0$ for $1 \le j \le 4$ and

$$E^{(5)}(0) = 12 \cdot \mathrm{Re}[2\pi i c^5 \epsilon^4 R_m^2 \gamma] + o(\epsilon^4),$$

where the contribution from the last complex component is

$$\begin{aligned}
&(m - n)^2 (m - 2p - 1)^2 + 4(m - n)^2 (m - 2p - 1)(m - k - p) \\
&\quad - 8(n - p)(m - p)(m - n)(m - k - p) \\
&\quad - 4(m - n)(m - 2p - 1)[(n - p)(m - p + 1) + (m - p)(2n - p - k + 1)].
\end{aligned}$$

We must add to this the contribution of the first complex components arising from the term $\hat{Z}_{ttw}(0) \cdot \hat{Z}_{tw}(0)\phi_t(0)$ which is

$$\begin{aligned}
&4(m - n)[2(2m - n - p - k)(n - p)^2 + (2m - n - 2p - 1)(n - p)(n + 1 - k) \\
&\quad + (2m - n - 2p - 1)(n - p)(2n - p - k + 1)].
\end{aligned}$$

It follows that $\gamma > 0$. Thus one can make $E^{(5)}(0) < 0$ for a suitable choice of c. □

Let us return to the case (C_4) : $3n \le 2m - 2 < 4n$ which splits into the two subcases $2m - 2 \equiv 0 \bmod 4$ and $2m - 2 \equiv 2 \bmod 4$. The first one was dealt with by $E^{(4)}(0)$, cf. Theorem 2.2, the second by $E^{(5)}(0)$, see Theorem 2.4. Combining both results we obtain

Theorem 2.5 *Let* $\hat{X}$ *be a minimal surface in normal form having the branch point* $w = 0$ *with the order* n *and the index* m *such that* (C_4) *holds. Then* $\hat{X}$ *cannot be a weak minimizer of* D *in* $\mathcal{C}(\Gamma)$.

We want to show how to calculate the fourth derivative assuming

$$2m + 2 = 3(n + 1) + r, \quad 1 \le r \le n. \tag{2.54}$$

The new approach consists in choosing the generator $\tau = \phi(0)$ as

$$\begin{aligned}
&\tau = \tau_0 + \tau_1 \quad \text{with } \tau_0 := \epsilon c w^{-n-1} + \epsilon \overline{c} w^{n+1}, \\
&\tau_1 := c w^{-r} + \overline{c} w^r, \quad c \in \mathbb{C}.
\end{aligned} \tag{2.55}$$

We need the following auxiliary result:

Lemma 2.5 *For any $\nu \in \mathbb{N}$ and $a \in \mathbb{C}$ we have*

$$\{2H[\mathrm{Re}(aw^{-\nu})]\}_w = \nu \overline{a} w^{\nu-1} \quad \text{on } \overline{B}. \tag{2.56}$$

Proof On S^1 one has $w^{-\nu} = \overline{w}^{\nu}$ whence

$$aw^{-\nu} = a\overline{w}^{\nu} = \overline{\overline{a}w^{\nu}} \quad \text{on } S^1$$

and therefore

$$\mathrm{Re}(aw^{-\nu}) = \mathrm{Re}(\overline{a}w^{\nu}) \quad \text{on } S^1.$$

Consequently

$$2H[\mathrm{Re}(aw^{-\nu})] = 2H[\mathrm{Re}(\overline{a}w^{\nu})] \quad \text{on } \overline{B}.$$

This implies

$$\{2H[\mathrm{Re}(aw^{-\nu})]\}_w = \{2H[\mathrm{Re}(\overline{a}w^{\nu})]\}_w \quad \text{on } \overline{B}.$$

Finally, since $\overline{a}w^{\nu}$ is holomorphic in $\mathbb{C}$, it follows that

$$\{2H[\mathrm{Re}(\overline{a}w^{\nu})]\}_w = \frac{d}{dw}(\overline{a}w^{\nu}) = \nu\overline{a}w^{\nu-1} \quad \text{on } \overline{B}. \qquad \square$$

Now we calculate $E^{(4)}(0)$ using the formulae (2.37) and (2.39):

$$\begin{aligned} E^{(4)}(0) = {} & 12\,\mathrm{Re}\int_{S^1} \hat{Z}_{ttw}(0)\cdot[w\hat{Z}_{tw}(0)\tau + w\hat{X}_w\phi_t(0)]\,dw \\ & + 12\,\mathrm{Re}\int_{S^1} w\hat{Z}_{tw}(0)\cdot\hat{Z}_{tw}(0)\phi_t(0)\,dw. \end{aligned} \tag{2.57}$$

From

$$w\hat{X}_w = (A_1 w^{n+1} + \cdots + A_{2m-2n+1}w^{2m-n+1} + \cdots, R_m w^{m+1} + \cdots)$$

it follows that

$$\begin{aligned} w\hat{X}_w\tau = {} & c\epsilon(A_1 + \cdots + A_{2m-2n+1}w^{2m-2n} + \cdots, R_m w^{m-n} + \cdots) \\ & + c(A_1 w^{n+1-r} + \cdots + A_{2m-2n+1}w^{2m-n-r+1} + \cdots, R_m w^{m+1-r} + \cdots) \\ & + g(w), \quad g(w) := w\hat{X}_w(w)\cdot[\epsilon\overline{c}w^{n+1} + \overline{c}w^{r}]. \end{aligned}$$

The expression $g(w)$ is "better" than the sum $T_1 + T_2$ of the first two terms T_1, T_2 on the right-hand side of this equation, in the sense that it is built in a similar way as $T_1 + T_2$ except that it is less singular. In the sequel this phenomenon will appear repeatedly, and so we shall always use a notation similar to the following:

$$w\hat{X}_w\tau = T_1 + T_2 + \langle\text{better}\rangle.$$

This sloppy notation will not do any harm since in the end we shall see that each of the two integrands in (2.57) possesses exactly one term of order w^{-1} as w-terms of least order, and no expression labelled "better" is contributing to them.

Using (2.35) one obtains

$$\begin{aligned}\hat{Z}_{tw}(0) = ic\epsilon(A_2 + \cdots + (2m-2n)A_{2m-2n+1}w^{2m-2n-1} + \cdots,\\ (m-n)R_m w^{m-n-1} + \cdots)\\ + ic((n+1-r)A_1 w^{n-r} + \cdots\\ + (2m-n+1-r)A_{2m-2n+1}w^{2m-n-r} + \cdots,\\ (m+1-r)R_m w^{m-r} + \cdots) + \langle\text{better}\rangle.\end{aligned}$$

This implies

$$\begin{aligned}w\hat{Z}_{tw}(0)\tau = ic^2\epsilon^2(A_2 w^{-n} + \cdots + (2m-2n)A_{2m-2n+1}w^{2m-3n-1} + \cdots,\\ (m-n)R_m w^{m-2n-1} + \cdots)\\ + ic^2\epsilon((n+1-r)A_1 w^{-r} + \cdots\\ + (2m-n+1-r)A_{2m-2n+1}w^{2m-2n-r} + \cdots,\\ (m+1-r)R_m w^{m-n-r} + \cdots) + \langle\text{better}\rangle.\end{aligned}$$

Recall that $A_k = \lambda_k A_1$ for $k = 1, \ldots, 2m-2n$. In order to remove all poles in the first two components of

$$f := w\hat{Z}_{tw}(0)\tau + w\hat{X}_w\phi_t(0)$$

one chooses $\phi_t(0)$ in a fashion similar to that used in the proof of Theorem 2.2:

$$\phi_t(0) := -ic^2\lambda_2\epsilon^2 w^{-2n-1} - ic^2\epsilon(n+1-r)w^{-n-1-r} + \cdots.$$

Then

$$\begin{aligned}f = ic^2\epsilon^2(\ldots(2m-2n)A_{2m-2n+1}w^{2m-3n-1} + \cdots, (m-n)R_m w^{m-2n-1} + \cdots)\\ + ic^2\epsilon(\ldots(2m-n+1-r)A_{2m-2n+1}w^{2m-2n-r} + \cdots,\\ 2(m-n)R_m w^{m-n-r} + \cdots) + \langle\text{better}\rangle.\end{aligned}$$

Here and in the sequel, $\ldots$ stand for non-pole terms with coefficients A_j with $j \le 2m-2n$.

The first two components of f (i.e. the expressions before the commata) are holomorphic; the worst pole in the third component is the term with the power w^{m-2n-1}; note that

$$\gamma := m - 2n - 1 = \frac{1}{2}[(2m+2) - 4(n+1)] < 0.$$

Thus Lemma 2.5 yields

$$\{H[\text{Re}(R_m w^\gamma)]\}_w = -\gamma\overline{R}_m w^{-\gamma-1}.$$

Using a formula established in the proof of Lemma 2.4 one obtains

$$\begin{aligned}\hat{Z}_{ttw}(0) = -c^2\epsilon^2(\ldots(2m-2n)(2m-3n-1)A_{2m-2n+1}w^{2m-3n-2},\\ (m-n)(2n+1-m)\overline{R}_m w^{2n-m} + \cdots)\\ - c^2\epsilon(\ldots(2m-n)(2m-2n-r)A_{2m-2n+1}w^{2m-2n-r-1} + \cdots,\\ (m-n)(m-n-r)R_m w^{m-n-r-1}) + \langle\text{better}\rangle.\end{aligned}$$

It follows that

$$\hat{Z}_{ttw}(0) \cdot [w\hat{Z}_{tw}(0)\tau + w\hat{X}_w\phi_t(0)] \\ = \{-ic^4\epsilon^3(m-n)^2(m-n-r)R_m^2 w^{-1} + \cdots\} + o(\epsilon^3)$$

since

$$2m - 3n - r - 2 = (2m+2) - [3(n+1) + r] - 1 = -1.$$

A straightforward calculation shows

$$w\hat{Z}_{tw}(0) \cdot \hat{Z}_{tw}(0)\phi_t(0) \\ = \{ic^4\epsilon^3(m-n)^2(n+1-r)R_m^2 w^{-1} + \cdots\} + o(\epsilon^3).$$

Thus one obtains by (2.57) that

$$E^{(4)}(0) = 12\epsilon^3 \operatorname{Re} \int_{S^1} ikc^4 R_m^2 \frac{dw}{w} + o(\epsilon^3)$$

with

$$k := (m-n)^2(n+1-r) - 2(m-n)^2(m-n-r).$$

Since

$$m - n - r = \frac{1}{2}\{(2m+2) - 2(n+1) - 2r\} = \frac{1}{2}(n+1-r)$$

it follows that $k = 0$.

This shows us that the leading term of some derivatives, may in fact, be zero. We conclude this section with a formula for the fifth derivative (the calculation of which we leave as an exercise for the reader), assuming

$$2m + 2 = 4(n+1) + r$$

where

$$4n + 4 < 2m + 2 \leq 5(n+1).$$

Setting our generator $\tau := \epsilon c/z^{n+1} + c/z^r$ we obtain

$$E^5(0) = 12 \operatorname{Re}[2\pi i c^5 \epsilon^4 \gamma R_m^2] + O(\epsilon^5) \tag{2.58}$$

where

$$\gamma = 5(m-n)^2(m-2n-1)^2 > 0.$$

We want to show that it often is possible to estimate the index m of an interior branch point w_0 of a minimal surface $\hat{X} \in \mathcal{C}(\Gamma)$ with the aid of a geometric condition on its boundary contour Γ. Following an idea by J.C.C. Nitsche, we use Radó's lemma for this purpose (Dierkes, Hildebrandt and Sauvigny [1] Sect. 4.9), which states the following. *If $f \in C^0(\overline{B})$ is harmonic in B, $f(w) \not\equiv 0$ in B, and $\nabla^j f(w_0) = 0$ at $w_0 \in B$ for $j = 0, 1, \ldots, m$, then f has at least $2(m+1)$ different zeros on ∂B.*

We can assume that the minimal surface $\hat{X}$ is transformed into the normal form with respect to the branch point $w_0 = 0$ having the index m. If the contour Γ is nonplanar, then $X^3(w) \not\equiv X_0^3 := X^3(0)$, whence $m < \infty$ and

$$X^3(w) = X_0^3 + \mathrm{Re}[cw^{m+1} + O(w^{m+2})] \quad \text{for } w \to 0$$

with $c \in \mathbb{C} \setminus \{0\}$. Hence $f := X^3 - X_0^3$ satisfies the assumptions of Radó's lemma, and therefore f has at least $2(m+1)$ different zeros on ∂B. Hence the plane $\Pi := \{(x^1, x^2, x^3) \in \mathbb{R}^3 : x^3 = X_0^3\}$ intersects Γ in at least $2(m+1)$ different points. If $m = \infty$ then even $\Gamma \subset \Pi$, and so we obtain:

Proposition 2.7 *If the minimal surface $\hat{X} \in \mathcal{C}(\Gamma)$ possesses a branch point $w_0 \in B$ with the index m, then there is a plane Π in $\mathbb{R}^3$ which intersects Γ in at least $2(m+1)$ different points. Consequently, if every plane in $\mathbb{R}^3$ intersects Γ in at most k different points, then the index m is bounded by*

$$2m + 2 \le k.$$

This result motivates the following

Definition 2.4 The **cut number** $c(\Gamma)$ of a closed Jordan curve Γ in $\mathbb{R}^3$ is the supremum of the number of intersection points of Γ with any (affine) plane Π in $\mathbb{R}^3$, i.e.

$$c(\Gamma) := \sup\{\sharp(\Gamma \cap \Pi) : \Pi = \text{affine plane in } \mathbb{R}^3\}. \tag{2.59}$$

It is easy to see that

$$4 \le c(\Gamma) \le \infty, \tag{2.60}$$

and for any nonplanar, real analytic, closed Jordan curve the cut number $c(\Gamma)$ is finite.

We can rephrase the second statement of Proposition 2.7 as follows:

Proposition 2.8 *The index m of any interior branch point of a minimal surface $\hat{X} \in \mathcal{C}(\Gamma)$ is bounded by*

$$2m + 2 \le c(\Gamma). \tag{2.61}$$

If n is the order and m the index of some branch point, then $1 \le n < m$. On the other hand, $c(\Gamma) = 4$ implies $m \le 1$, and $c(\Gamma) = 6$ yields $m \le 2$. Thus we obtain

Corollary 2.1

(i) *If $c(\Gamma) = 4$ then every minimal surface $\hat{X} \in \mathcal{C}(\Gamma)$ is free of interior branch points.*
(ii) *If $c(\Gamma) = 6$ then any minimal surface $\hat{X} \in \mathcal{C}(\Gamma)$ has at most simple interior branch points of index two; if $\hat{X}$ has an interior branch point, it cannot be a weak minimizer of D in $\mathcal{C}(\Gamma)$.*

Proof (i) follows from $1 \le n < m \le 1$, which is impossible. (ii) $1 \le n < m \le 2$ implies $n = 1$ and $m = 2$ for an interior branch point w_0 of $\hat{X}$, whence $2n \le 2m - 2 < 3$. Thus condition (C_3) is satisfied, and therefore the last assertion follows from Theorem 2.1. □

Corollary 2.2 *Let $\hat{X} \in \mathcal{C}(\Gamma)$ be a minimal surface with an interior branch point of order n, and suppose that the cut number of Γ satisfies $c(\Gamma) \le 4n + 3$. Then $\hat{X}$ is not a weak minimizer of D in $\mathcal{C}(\Gamma)$.*

Proof By (2.61) we have

$$2m + 2 \le 4n + 3;$$

hence either

$$2n + 4 \le 2m + 2 < 3n + 4 \Leftrightarrow 2n \le 2m - 2 < 3n$$

or

$$3n + 4 \le 2m + 2 < 4n + 4 \Leftrightarrow 3n \le 2m - 2 < 4n$$

hold true, i.e. either (C_3) or (C_4) is fulfilled. In the first case the assertion follows from Theorem 2.1, in the second from Theorem 2.5. □

Finally, we mention Catalan's surface (picture on our cover) which has a branch point at $w = 0$, with $n = 1$ and $m = 2$. Thus, $2m - 2 < 3n$, and so Wienholtz's theorem applies. The normal form for $\hat{X}_w$ is

$$\hat{X}_w = \left(\frac{i}{2}(e^z - e^{-z}), e^{\frac{z}{2}} - e^{-\frac{z}{2}}, 1 - \frac{1}{2}(e^z + e^{-z})\right).$$

Summary

In this section we have calculated derivatives of Dirichlet's energy with respect to various generators τ. In general it will be extremely difficult to calculate higher order derivatives using an arbitrary choice of generators. Remarkably, with the appropriate choice of generators, the higher order derivatives can be simply calculated and the results are independent of the first complex components of $\hat{X}_w$. We show this in the next three chapters.

Chapter 3
Very Special Case; The Theorem for $n + 1$ Even and $m + 1$ Odd

In this chapter we want to show that *a (nonplanar) weak relative minimizer $\hat{X}$ of Dirichlet's integral D that is given in the normal form cannot have $w = 0$ as a branch point if its order n is odd and its index m is even.* Note that such a branch point is not exceptional since $n + 1$ cannot be a divisor of $m + 1$. We shall give the proof only under the assumptions $n \geq 3$ since $n = 1$ is easily dealt with by a method presented in the next section. (Moreover it would suffice to treat the case $m \geq 6$ since $2m - 2 < 3n$ is already treated by the Wienholtz theorem. So $2m \geq 3n + 2 \geq 11$, i.e. $m \geq 6$ since m is even.)

3.1 The Strategy of the Proof

The strategy to find the first non-vanishing derivative of $E(t)$ at $t = 0$ that can be made negative consists in the following four steps:

(I) Guess the candidate L for which $E^{(L)}(0) < 0$ can be achieved with a suitable choice of the generator $\tau = \phi(0)$.

(II) Select $D_t^\beta \phi(0), \beta \geq 1$, so that the lower order derivatives $E^{(j)}(0), j = 1, 2, \ldots, L-1$ vanish, ($D_t^\beta := \frac{\partial^\beta}{\partial t^\beta}$).

(III) Prove that

$$E^{(L)}(0) = \operatorname{Re} \int_{S^1} c^L k R_m^2 \frac{dw}{w} = \operatorname{Re}\{2\pi i c^L k R_m^2\}$$

where $c \neq 0$ is a complex number which can be chosen arbitrarily, and $k \in \mathbb{C}$ is to be computed.

(IV) Show that $k \neq 0$.

Remark 3.1 In order to achieve (II) one tries to choose $D_t^\beta \phi(0), \beta \geq 1$, in such a way that the integrands of $E^{(j)}(0)$ for $j < L$ are free of any poles and, therefore, free of first-order poles. To see that this strategy is advisable, let us consider the

A. Tromba, *A Theory of Branched Minimal Surfaces*,
Springer Monographs in Mathematics,
DOI 10.1007/978-3-642-25620-2_3, © Springer-Verlag Berlin Heidelberg 2012

case $L=5$; then we have to achieve $E^{(4)}(0)=0$. Recall that $E^{(4)}(0)$ consists of two terms, one of which has the form

$$I := 12\,\mathrm{Re}\int_{S^1}\{2H[\mathrm{Re}\, if]\}_w f\,dw$$

where

$$f := w[iw\hat{X}_w\tau]_w\tau + w\hat{X}_w\phi_t(0).$$

Assume that f had poles, say,

$$f(w)=g(w)+h(w), \quad g(w)=\sum_{j\ge 1}a_j w^{-j}, \; h = \text{ holomorphic in } B,$$

and $h\in C^0(\overline{B})$. Then, by Lemma 2.5,

$$\{2H[\mathrm{Re}\, if]\}_w(w)=g^*(w)+h'(w), \quad g^*(w):=-i\sum_{j\ge 1} j\overline{a}_j w^{j-1}.$$

Thus, $I=12\cdot\{I_1+I_2+I_3\}$, with

$$I_1 := \mathrm{Re}\int_{S^1} g^* g\,dw, \qquad I_2 := \mathrm{Re}\int_{S^1} h' g\,dw, \qquad I_3 := \mathrm{Re}\int_{S^1}(g^*h+h'h)\,dw.$$

The worst term is I_1; one obtains

$$I_1=\mathrm{Re}\int_{S^1}\sum_{j,\ell\ge 1}(-ij\overline{a}_j w^{j-1}a_\ell w^{-\ell})\,dw = 2\pi\sum_{j\ge 1} j|a_j|^2>0$$

and $I_3=0$. Hence, in order to achieve $I=0$, one would have to balance I_2 against $I_1>0$ which seems to be pretty hopeless.

Let us now apply the "strategy" to prove

Theorem 3.1 *Let $\hat{X}$ be a nonplanar minimal surface in normal form that has $w=0$ as a branch point of odd order $n\ge 3$ and of even index $m\ge 4$. Then, by a suitable choice of $\tau=\phi(0)$ and $D_t^\beta\phi(0)$, one can achieve that*

$$E^{(m+1)}(0)<0 \quad \text{and} \quad E^{(j)}(0)=0 \quad \text{for } 1\le j\le m.$$

Proof Set $N:=L-1$, $M:=L-(\alpha+\beta+1)=N-(\alpha+\beta)$, hence $L-1=\alpha+\beta+M$. By Leibniz's formula,

$$D_t^N\{[\hat{Z}_w\cdot\hat{Z}_w]\phi\}=\sum_{\alpha=0}^{N-\beta}\sum_{\beta=0}^{N}\frac{N!}{\alpha!\beta!(N-\beta-\alpha)!}(D_t^{N-\beta-\alpha}\hat{Z}_w)\cdot(D_t^\alpha\hat{Z}_w)D_t^\beta\phi.$$

Since

$$D_tE(t)=2\,\mathrm{Re}\int_{S^1} w\hat{Z}(t)_w\cdot\hat{Z}(t)_w\,\phi(t)\,dw,$$

we can use Leibniz's formula to compute $E^{(L)}(t)$ from

$$E^{(L)}(t) = 2\,\mathrm{Re} \int_{S^1} w D_t^N \{[\hat{Z}_w(t) \cdot \hat{Z}_w(t)]\phi(t)\}\, dw.$$

We choose $L := m + 1$; then $L \geq 5$ as we have assumed $m \geq 4$. It follows that

$$E^{(L)}(0) = J_1 + J_2 + J_3 \tag{3.1}$$

where the terms J_1, J_2, J_3 are defined as follows: Set

$$T^{\alpha,\beta} := w(D_t^\alpha \hat{Z}(0))_w D_t^\beta \phi(0). \tag{3.2}$$

Then,

$$\begin{aligned}
J_1 :=\ & 4\,\mathrm{Re} \int_{S^1} [D_t^{L-1}\hat{Z}(0)]_w \cdot (w\hat{X}_w \tau)\, dw \\
& + 4 \cdot (L-1)\,\mathrm{Re} \int_{S^1} [D_t^{L-2}\hat{Z}(0)]_w f\, dw \\
& + 4 \sum_{M > \frac{1}{2}(L-1)}^{L-3} \frac{(L-1)!}{M!(L-M-1)!}\,\mathrm{Re} \int_{S^1} [D_t^M \hat{Z}(0)]_w \cdot g_{L-M-1}\, dw,
\end{aligned} \tag{3.3}$$

$$f := T^{1,0} + T^{0,1} = w[\hat{Z}_t(0)]_w \tau + w\hat{X}_w \phi_t(0),$$

$$g_\nu := \sum_{\alpha+\beta=\nu} c_{\alpha\beta}^\nu T^{\alpha,\beta} \quad \text{with } c_{\alpha\beta}^\nu := \frac{\nu!}{\alpha!\beta!},\ \alpha + \beta + \nu = L - 1;$$

$$\begin{aligned}
J_2 :=\ & \sum_{M=2}^{\frac{1}{2}(L-1)} \frac{2(L-1)!}{M!M!}\,\mathrm{Re} \int_{S^1} [D_t^M \hat{Z}(0)]_w \cdot h_M\, dw \\
& + 2(L-1)(L-2)\,\mathrm{Re} \int_{S^1} [\hat{Z}_t(0)]_w \cdot T^{1,L-3}\, dw,
\end{aligned} \tag{3.4}$$

$$h_M := \sum_{\alpha=0}^{M} \psi(M,\alpha) \frac{M!}{\alpha!(L-1-M-\alpha)!} T^{\alpha,L-1-M-\alpha},$$

$$\psi(M,\alpha) := 1 \quad \text{for } \alpha = M, \qquad \psi(M,\alpha) := 2 \quad \text{for } \alpha \neq M;$$

$$\begin{aligned}
J_3 :=\ & 4(L-1)\,\mathrm{Re} \int_{S^1} w\hat{Z}_{tw}(0) \cdot \hat{X}_w D_t^{L-2}\phi(0)\, dw \\
& + 2\,\mathrm{Re} \int_{S^1} w\hat{X}_w \cdot \hat{X}_w D_t^{L-1}\phi(0)\, dw.
\end{aligned} \tag{3.5}$$

We have $J_3 = 0$ since $\hat{X}_w \cdot \hat{X}_w = 0$ and $\hat{Z}_{tw}(0) \cdot \hat{X}_w = 0$ on account of formula (36) in Dierkes, Hildebrandt and Tromba [1], Chapter 6.1.

Now we proceed as follows:

Step 1. We choose $\tau = \phi(0)$ and $D_t^\beta \phi(0)$ for $\beta \geq 1$ in such a way that f and g_{L-M-1} are holomorphic. Then the integrands of the three integrals in J_1 are holomorphic because all w-derivatives $[D_t^j \hat{Z}(0)]_w$ of the harmonic functions $D_t^j \hat{Z}(t)$ are holomorphic. Then it follows that $J_1 = 0$, and thus we have

$$E^{(L)}(0) = J_2. \tag{3.6}$$

Step 2. Then it will be shown that $E^{(L)}(0)$ reduces to the single term

$$E^{(L)}(0) = \frac{2 \cdot m!}{(\frac{m}{2})!\,(\frac{m}{2})!} \operatorname{Re} \int_{S^1} w[D_t^{m/2} \hat{Z}(0)]_w \cdot [D_t^{m/2} \hat{Z}(0)]_w \tau \, dw \tag{3.7}$$

which can be calculated explicitly; it will be shown that

$$E^{(L)}(0) = \frac{2 \cdot m!}{(\frac{m}{2})!\,(\frac{m}{2})!} \operatorname{Re}(2\pi i \cdot \kappa \cdot R_m^2) \tag{3.8}$$

where κ is the number

$$\kappa := i^{L-1}(a - ib)^L (m-1)^2 (m-3)^2 \ldots 3^2 \cdot 1^2 \tag{3.9}$$

if the generator $\tau = \phi(0)$ is chosen as

$$\tau(w) := (a - ib)w^{-2} + (a + ib)w^2. \tag{3.10}$$

For a suitable choice of $(a - ib)$ one obtains $E^{(L)}(0) < 0$. Furthermore the construction will yield $E^{(j)}(0) = 0$ for $1 \le j \le L-1$.

Before we carry out this program for general $n \ge 3$, $m \ge 4$, $n =$ odd, $m =$ even, we explain the procedure for the simplest possible case: $n = 3$ and $m = 4$.

From the normal form for $\hat{X}_w$ with the order n and the index m of the branch point $w = 0$ we obtain

$$w\hat{X}_w = (A_1 w^{n+1} + \cdots + A_{2m-2n+1} w^{2m-n+1} + \cdots, R_m w^{m+1} + \cdots). \tag{3.11}$$

Choosing τ according to (3.10) it follows from

$$[\hat{Z}_t(0)]_w = (iw\hat{X}_w\tau)_w$$

that

$$\begin{aligned} [\hat{Z}_t(0)]_w = {} & (a - ib)(i(n-1)A_1 w^{n-2} + inA_2 w^{n-1} + \cdots \\ & + i(2m-n-1)A_{2m-2n+1} w^{2m-n-2}, i(m-1)R_m w^{m-2} + \cdots) \\ & + \langle \text{better} \rangle. \end{aligned} \tag{3.12}$$

Here, $\langle$better$\rangle$ stands again for terms that are similarly built as those in the preceding expression but whose w-powers attached to corresponding coefficients are of higher order. Then

$$\begin{aligned} w[\hat{Z}_t(0)]_w \tau = {} & (a - ib)^2 (i(n-1)A_1 w^{n-3} + inA_2 w^{n-2} + \cdots \\ & + i(2m-n-1)A_{2m-2n+1} w^{2m-n-3} + \cdots, i(m-1)R_m w^{m-3} + \cdots) \\ & + \langle \text{better} \rangle. \end{aligned} \tag{3.13}$$

Since this term is holomorphic we have the freedom to set $\phi_t(0) = 0$. Then $f(w) = w\hat{Z}_{tw}(0)\tau + w\hat{X}_w\phi_t(0)$ is holomorphic, and Proposition 2.6 in Chap. 2 yields

$$E^{(5)}(0) = 12\,\mathrm{Re}\int_{S^1} w\hat{Z}_{ttw}(0)\cdot\hat{Z}_{ttw}(0)\tau\,dw. \tag{3.14}$$

(This follows of course also from the general formulae stated above.)

From formula (2.51) following Lemma 2.4 we get

$$\hat{Z}_{ttw}(0) = \{iw[iw\hat{X}_w\tau]_w\tau\}_w = i\{w\hat{Z}_{tw}(0)\tau\}_w,$$

and so

$$\begin{aligned}\hat{Z}_{ttw}(0) = -(a-ib)^2((n-1)(n-3)A_1w^{n-4} + \cdots \\ + (2m-n-1)(2m-n-3)A_{2m-2n+1}w^{2m-n-4} + \cdots, \\ (m-1)(m-3)R_mw^{m-4} + \cdots) + \langle\text{better}\rangle.\end{aligned} \tag{3.15}$$

Since $n-3=0$ and $m=4$, this leads to

$$\hat{Z}_{ttw}(0)\cdot\hat{Z}_{ttw}(0) = (a-ib)^4(m-1)^2(m-3)^2R_m^2 + \cdots, \tag{3.16}$$

and by (3.14) we obtain for $L = m+1 = 5$:

$$\begin{aligned}E^{(L)}(0) = E^{(5)}(0) &= 12\cdot\mathrm{Re}\int_{S^1}(a-ib)^5(m-1)^2(m-3)^2R_m^2\frac{dw}{w} \\ &= 12\cdot\mathrm{Re}[2\pi i(a-ib)^5(m-1)^2(m-3)^2R_m^2], \quad m=4.\end{aligned} \tag{3.17}$$

Now we turn to the **general case** of an odd $n \ge 3$ and an even index $m \ge 4$.

Step 1. *The pole-removal technique to make the expressions f and g_{L-M-1} in the integral J_1 holomorphic.*

We have already seen that $f(w)$ is holomorphic if we set $\phi_t(0) = 0$. In fact, we set

$$D_t^\beta\phi(0) = 0 \quad \text{for } 1 \le \beta \le \frac{n-1}{2} \text{ and for } \beta > \frac{1}{2}(L-3) \tag{3.18}$$

and prove the following

Lemma 3.1 *By the pole-removal technique we can inductively choose $D_t^\beta\phi(0)$ for $\beta \le \frac{1}{2}(L-3)$ such that g_ν is holomorphic for $\nu = 0, 1, \ldots, \frac{1}{2}(L-3)$. Then the derivative $[D_t^\gamma\hat{Z}(0)]_w$ is not only holomorphic, but can be obtained in the form*

$$[D_t^\gamma\hat{Z}(0)]_w = \{ig_{\gamma-1}\}_w \quad \textit{for } \gamma = 1, 2, \ldots, \frac{1}{2}(L-1). \tag{3.19}$$

Suppose this result were proved. Since in J_1 there appear only g_ν with $\nu = L - M - 1$ where $\frac{1}{2}(L-1) < M \le L-3$, i.e. $2 \le \nu \le \frac{1}{2}(L-3)$, all integrands in J_1 were indeed holomorphic, and so $J_1 = 0$. Thus it remains to prove Lemma 3.1.

Proof (*of Lemma* 3.1) By definition we have

$$g_\nu = \sum_{\alpha+\beta=\nu} c^\nu_{\alpha\beta} T^{\alpha,\beta}, \quad T^{\alpha,\beta} := w[D^\alpha_t \hat{Z}(0)]_w D^\beta_t \phi(0), \tag{3.20}$$

and $\phi(0) = \tau$.

The expressions $w[D^\alpha_t \hat{Z}(0)]_w \tau$ have no pole for $\alpha \leq \frac{n-1}{2}$, and we make the important observation that there are numbers c, c' such that

$$w[D_t^{\frac{n-1}{2}} \hat{Z}(0)]_w \tau = (cA_1 + \cdots, c' R_m w^{m-n} + \cdots).$$

Thus, a pole in $w[D^\alpha_t \hat{Z}(0)]_w \tau$ may arise at first for $\alpha = \frac{1}{2}(n+1)$; then we have, say

$$w[D_t^{\frac{1}{2}(n+1)} \hat{Z}(0)]_w \tau = (cA_2 w^{-1} + \cdots, c' R_m w^{m-n-2} + \cdots). \tag{3.21}$$

This requires a non-zero $D_t^{\frac{n+1}{2}} \phi(0)$ in case that $cA_2 \neq 0$ if we want to make $g_{\frac{1}{2}(n+1)}$ pole-free. Now we go on and discuss the pole removal for $\nu = \frac{1}{2}(n+3), \frac{1}{2}(n+5), \ldots, \frac{1}{2}(L-3)$.

Observation 3.1 Since m is even, n is odd, and $m > n$, we have

$$m = n + (2k+1), \quad k = 0, 1, 2, \ldots, \tag{3.22}$$

and therefore

$$\frac{1}{2}(L-3) = \frac{1}{2}(m-2) = \frac{1}{2}(n+2k-1). \tag{3.23}$$

Thus, for $m = n+1$, all g_ν with $2 \leq \nu \leq \frac{1}{2}(L-3)$ are pole-free if we set $D^\beta_t \phi(0) = 0$ for all $\beta \geq 1$; cf. (3.18). For $m = n+3$, we have to choose $D^\beta_t \phi(0)$ appropriately for $\beta = \frac{1}{2}(n+1)$ while the other $D^\beta_t \phi(0)$ are taken to be zero. For $m = n+5$, we must also choose $D^\beta_t \phi(0)$ appropriately for $\beta = \frac{1}{2}(n+3)$ whereas the other $D^\beta_t \phi(0)$ are set to be zero. In this way we proceed inductively and choose $D^\beta_t \phi(0)$ in a suitable way for $\beta = \frac{1}{2}(n+1), \frac{1}{2}(n+3), \ldots, \frac{1}{2}(n+2k-1)$ in case that $m = n+2k+1$ while all other $D^\beta_t \phi(0)$ are taken to be zero according to (3.18).

Observation 3.2 The pole-removal procedure would only stop for some g_ν with $\frac{1}{2}(n+1) \leq \nu \leq \frac{1}{2}(L-3)$ if the w-power attached to $A_{2m-2n+1}$ became negative. We have to check that this does not happen for $\nu \leq \frac{1}{2}(L-3)$. Since at the α-th stage in defining $[D^\alpha_t \hat{Z}(0)]_w$ the w-powers have been reduced by 2α, we must check that the terms $T^{\alpha,\beta}$ have no poles connected with $A_{2m-2n+1}$ if $\alpha + \beta \leq \frac{1}{2}(L-3)$. Looking first only at $T^{\alpha,0} = w[D^\alpha_t \hat{Z}(0)]_w \tau$ for $\alpha \leq \frac{1}{2}(L-3)$, we must have

$$2m - n - 2\alpha = 2m - n + 1 - 2(\alpha+1) \geq 0 \quad \text{for } \alpha \leq \frac{1}{2}(L-3),$$

which is true since

$$2m - n + 1 - 2 \cdot \frac{1}{2}(L-1) = m - n + 1 > 0.$$

We must also check that during the process no pole is introduced into the third complex component. Again we first look at $T^{\alpha,0}$ for $\alpha \le \frac{1}{2}(L-3)$. Then the order of the w-power at the R_m-term is

$$m - 2\alpha - 1 = (m+1) - 2(\alpha+1) \ge (m+1) - (L-1) = 1,$$

and so there is no pole.

Let us now look at the pole-removal procedure. For $m = n+1$ all g_ν with $2 \le \nu \le \frac{1}{2}(L-3)$ are pole-free if we assume (3.18). If $m = n+3$ we have to make $g_{\frac{1}{2}(n+1)}$ pole-free. To this end it suffices to choose $D_t^{\frac{1}{2}(n+1)}\phi(0)$ appropriately; it need have a pole at most of order $(n+2)$ in order to remove a possible pole of $T^{\alpha,0}$, $\alpha = \frac{1}{2}(n+1)$, cf. (3.21).

If $m = n+5$, we have to choose $D_t^\beta \phi(0)$ appropriately for $\beta = \frac{1}{2}(n+1)$ and $\beta = \frac{1}{2}(n+3)$. The derivative $D_t^{\frac{1}{2}(n+1)}\phi(0)$ will be taken as before, while $D_t^{\frac{1}{2}(n+3)}\phi(0)$ is to be chosen in such a way that

$$g_{\frac{1}{2}(n+3)} = T^{\frac{1}{2}(n+3),0} + T^{1,\frac{1}{2}(n+1)} + T^{0,\frac{1}{2}(n+3)}$$

becomes holomorphic. Since

$$\begin{aligned} T^{1,\frac{1}{2}(n+1)} &= w[\hat{Z}_t(0)]_w D_t^{\frac{1}{2}(n+1)}\phi(0) \\ &= (i(n-1)(a-ib)A_1 w^{n-1} + \cdots, \\ &\qquad i(m-1)(a-ib)R_m w^{m-1} + \cdots) D_t^{\frac{1}{2}(n+1)}\phi(0) \\ &= (cA_1 w^{-3} + \cdots, c' R_m w^{m-n-3} + \cdots) \end{aligned}$$

with some constants c, c', the derivative $D_t^{\frac{1}{2}(n+3)}\phi(0)$ in

$$T^{0,\frac{1}{2}(n+3)} = w\hat{X}_w D_t^{\frac{1}{2}(n+3)}\phi(0)$$

should have a pole of order $n+4$, while a pole of lower order than $n+4$ is needed to remove a possible singularity in the first term $T^{\frac{1}{2}(n+3),0} = w[D_t^{\frac{1}{2}(n+3)}\hat{Z}(0)]_w \tau$.

In this way we can proceed inductively choosing the poles of $D_t^\beta \phi(0)$ always at most of order

$$n + 2\left(\beta - \frac{n-1}{2}\right) = 2\beta + 1 \quad \text{for } \frac{1}{2}(n+1) \le \beta \le \frac{1}{2}(L-3). \tag{3.24}$$

This is the crucial estimate on the order of the pole of $D_t^\beta \phi(0)$ in order to ensure that these derivatives play no role in the final calculations.

Observation 3.3 Consider the last complex component of

$$g_{\frac{1}{2}(n+1)} = w[D_t^{\frac{1}{2}(n+1)} \hat{Z}(0)]_w \tau + w \hat{X}_w D_t^{\frac{1}{2}(n+1)} \phi(0).$$

The lowest w-power attached to R_m in the first term is $1 + m - (n+1) - 2 = m - n - 2 \geq 1$ (since in this case $m \geq n+3$ according to Observation 3.1). The lowest w-power associated to R_m in the second term is $1 + m - (n+2) = m - n - 1 > m - n - 2$. Continuing inductively we see that the lowest w-power attached to R_m in any g_ν arises from $\tau = \phi(0)$ and not from any $D_t^\beta \phi(0)$. □

This ends the proof of Step 1, and we have found that $E^{(L)}(0) = J_2$. Now we come to

Step 2. The integral J_2 is a linear combination of the real parts of the integrals

$$I_{\alpha\gamma\beta} := \int_{S^1} w[D_t^\alpha \hat{Z}(0)]_w \cdot [D_t^\gamma \hat{Z}(0)]_w D_t^\beta \phi(0)\, dw \tag{3.25}$$

where $1 \leq \alpha, \gamma \leq \frac{1}{2}(L-1)$ and $\beta = (L-1) - \alpha - \gamma$. Then we have

$$\beta = 0 \quad \text{if and only if } \alpha = \gamma = \frac{1}{2}(L-1) = \frac{m}{2}. \tag{3.26}$$

This implies

$$J_2 = \frac{2 \cdot m!}{\left(\frac{m}{2}\right)! \left(\frac{m}{2}\right)!} \operatorname{Re} \int_{S^1} w[D_t^{\frac{m}{2}} \hat{Z}(0)]_w \cdot [D_t^{\frac{m}{2}} \hat{Z}(0)]_w \tau\, dw \tag{3.27}$$

because of the following

Lemma 3.2 *We have*

$$I_{\alpha\gamma\beta} = 0 \quad \textit{for } 1 \leq \alpha, \gamma \leq \frac{1}{2}(L-1) \quad \textit{and} \quad 1 \leq \beta = m - \alpha - \gamma. \tag{3.28}$$

Proof Let us first show that the product of the last complex components of $[D_t^\alpha \hat{Z}(0)]_w$ and $[D_t^\gamma \hat{Z}(0)]_w$ and of $w D_t^\beta \phi(0)$ have a zero integral. In fact, this product has the form

$$\begin{aligned}
&\text{const}(w R_m w^{m-2\alpha} \cdot R_m w^{m-2\gamma} + \cdots)(w^{-2\beta-1} + \cdots) \\
&\quad = \text{const}\, R_m^2 w^{1+2m-2(\alpha+\beta+\gamma)-1} + \cdots = \text{const} \cdot R_m^2 + \cdots
\end{aligned}$$

since $\alpha + \beta + \gamma = L - 1 = m$.

The same holds true for the scalar product of the first two complex components, multiplied by $w D_t^\beta \phi(0)$. To see this we assume without loss of generality that $\alpha \geq \gamma$. Denote by $P^{\alpha\gamma}$ the expression

$$P^{\alpha\gamma} := w[C_1^\alpha \cdot C_1^\gamma + C_2^\alpha \cdot C_2^\gamma]$$

where C_1^α, C_2^α and C_1^γ, C_2^γ are the first two complex components of $[D_t^\alpha \hat{Z}(0)]_w$ and $[D_t^\gamma \hat{Z}(0)]_w$ respectively.

Case 1. If $2\gamma \leq 2\alpha < n$ then

$$P^{\alpha\gamma} = w(\mathrm{const}\, A_j w^{n-2\alpha} + \cdots + \mathrm{const}\, A_{2m-2n+1} w^{2m-n-2\alpha} + \cdots)$$
$$\cdot (\mathrm{const}\, A_\ell w^{n-2\gamma} + \cdots + \mathrm{const}\, A_{2m-2n+1} w^{2m-n+\gamma} + \cdots)$$

with $j, \ell < 2m - 2n + 1$.

Case 2. If $2\gamma < n < 2\alpha$ then

$$P^{\alpha\gamma} = w(\mathrm{const}\, A_j + \cdots + \mathrm{const}\, A_{2m-2n+1} w^{2m-n} + \cdots)$$
$$\cdot (\mathrm{const}\, A_\ell w^{n-2\gamma} + \cdots + \mathrm{const}\, A_{2m-2n+1} w^{2m-n-2\gamma} + \cdots)$$

with $j, \ell < 2m - 2n + 1$.

Case 3. If $n < 2\alpha$ and $n < 2\gamma$ then

$$P^{\alpha\gamma} = w(\mathrm{const}\, A_j + \cdots + \mathrm{const}\, A_{2m-2n+1} w^{2m-n-2\alpha} + \cdots)$$
$$\cdot (\mathrm{const}\, A_\ell + \cdots + \mathrm{const}\, A_{2m-2n+1} w^{2m-n-2\gamma} + \cdots).$$

Let $\mu(\alpha, \gamma)$ be the lowest w-power appearing in $P^{\alpha\gamma} D_t^\beta \phi(0)$. Recalling $\alpha + \beta + \gamma = m$ we obtain the following results:

Case 1.

$$\begin{aligned}\mu(\alpha, \gamma) &= 1 + 2m - 2\gamma - 2\alpha - 2\beta - 1\\ &= 2 + 2m - 2(\alpha + \beta + \gamma + 1)\\ &= 2 + 2m - 2(m + 1) = 0.\end{aligned}$$

Case 2. $\mu(\alpha, \gamma)$ is either zero as in Case 1, or

$$\begin{aligned}\mu(\alpha, \gamma) &= 1 + 2m - n - 2\gamma - 2\beta - 1\\ &= 2 + 2m - n - 2(\gamma + \beta + 1)\\ &= 2 + 2m - n - 2(m + 1 - \alpha) = 2\alpha - n > 0.\end{aligned}$$

Case 3. As in Case 2 we have $\mu(\alpha, \gamma) > 0$.

This proves $I_{\alpha\gamma\beta} = 0$ for $1 \leq \alpha, \gamma \leq \frac{m}{2}$ and $1 \leq \beta = m - \alpha - \gamma$, which yields Lemma 3.2. □

Thus we have arrived at (3.27), and a straightforward computation leads to (3.8) and (3.9); so the proof of Theorem 3.1 is complete. □

Chapter 4
The First Main Theorem; Non-exceptional Branch Points; The Non-vanishing of the Lth Derivative of Dirichlet's Energy

Let us state our main goal: Assuming that $\hat{X} \in \mathcal{C}(\Gamma)$ is a nonplanar minimal surface in normal form having $w = 0$ as a branch point of order n and index m, we want to show that $\hat{X}$ cannot be a weak relative minimizer of Dirichlet's integral D in the class $\mathcal{C}(\Gamma)$. Unfortunately this goal cannot be achieved for all branch points but only for non-exceptional ones and special kinds of exceptional ones. In this chapter we investigate the non-exceptional branch points, while in Chaps. 5 and 6 we deal with the exceptional ones. The main result of the present section – our **First Main Theorem** – is the following

Theorem 4.1 *Let $\hat{X} \in C^\infty(\overline{B}, \mathbb{R}^3)$ be a nonplanar minimal surface in normal form having $w = 0$ as a non-exceptional branch point of order n and index m. Then $\hat{X}$ is not a weak relative minimizer of D.*

Recall that $w = 0$ is said to be **non-exceptional** if and only if

$$m + 1 \not\equiv 0 \bmod (n + 1).$$

According to Lemma 7.3 in Sect. 7.1 this is the case if either

$$(L - 1)(n + 1) < 2(m + 1) < L(n + 1) \quad \text{with an even } L \geq 4, \tag{4.1}$$

or

$$(L - 1)(n + 1) < 2(m + 1) \leq L(n + 1) \quad \text{with an odd } L \geq 3. \tag{4.2}$$

The strategy to prove Theorem 4.1 as outlined in Sects. 7.1 and 7.2 is to construct a variation $\hat{Z}(t)$, $|t| < t_0$, of $\hat{X}$ such that $E(t) := D(\hat{Z}(t))$ satisfies

$$\begin{aligned} &E^{(j)}(0) = 0 \quad \text{for } 1 \leq j \leq k-1 \quad \text{for some } k \geq 3 \quad \text{as well as} \\ &E^{(k)}(0) \neq 0 \quad \text{if } k \text{ is odd}, \quad \text{or} \quad E^{(k)}(0) < 0 \quad \text{if } k \text{ is even}, \end{aligned} \tag{4.3}$$

where

$$E^{(k)}(t) := \frac{d^k}{dt^k} E(t)$$

A. Tromba, *A Theory of Branched Minimal Surfaces*,
Springer Monographs in Mathematics,
DOI 10.1007/978-3-642-25620-2_4, © Springer-Verlag Berlin Heidelberg 2012

denotes the kth derivative of $E(t)$. Here $\hat{Z}(t)$ is defined as the harmonic extension of boundary values $Z(t)$ onto the disk $\overline{B}$, where $Z(\theta, t)$ are defined via the boundary values $X(\theta)$ of $\hat{X}$ by the formula $Z(\theta, t) := X(\gamma(\theta, t))$ with

$$\gamma(\theta, t) = \theta + \sigma(\theta, t)$$

where $\sigma \in C^\infty$ is 2π-periodic in θ and satisfies $\sigma(\theta, 0) = 0$. One obtains

$$\frac{d}{dt} E(t) = 2\,\mathrm{Re} \int_{S^1} w \hat{Z}(t)_w \cdot \hat{Z}(t)_w \phi(t)\, dw$$

with test functions $\phi(\theta, t)$ that are 2π-periodic in θ and such that $\phi_\nu(\theta) := D_t^\nu \phi(\theta, t)\big|_{t=0}$ can arbitrarily be chosen as C^∞-functions which are 2π-periodic. Then $E'(0) = 0$, and if $\phi(0)$ is chosen as the boundary value function $\tau|\partial B$ of the generator τ of an inner forced Jacobi field $\hat{h}$ attached to $\hat{X}$, i.e.

$$\phi(\theta, 0) = \tau(e^{i\theta}),$$

then also $E''(0) = 0$. As we have obtained in Chaps. 2 and 3, $\phi(0)$ will always be chosen in this way, and so the Laurent expansion of $\tau(w)$ is of the form

$$\tau(w) = \frac{c}{w^\ell} + \cdots \quad \text{with } \ell \le n + 1.$$

The derivatives

$$\phi_\nu = D_t^\nu \phi(\cdot, t)\big|_{t=0} = D_t^\nu \phi(0)$$

are appropriately chosen as boundary values of meromorphic functions such that the **pole-removal technique** (as explained in Chaps. 2 and 3) can be applied. With a slight misuse of notation we write τ and ϕ_ν both for the corresponding meromorphic functions and their boundary values. The trick in computing $E^{(j)}(0)$ consists in making as many terms of the integrand as possible to be boundary values of holomorphic functions. Consequently, their complex line integrals over S^1 vanish in virtue of Cauchy's integral theorem and (4.3) can be achieved for $k = L$.

This idea works very well in case (4.2) where L is odd, and we obtain

Theorem 4.2 *Suppose that $\hat{X}$ satisfies the assumptions of Theorem* 4.1 *while n, m fulfil condition* (4.2). *Then we can achieve* (4.3) *for $k = L$.*

The situation now is as follows: There is a well-defined Diophantine polynomial $p_L(x, y)$, called the **minimal surface polynomial of rank** L, which is independent of the specific minimal surface $\hat{X}$ satisfying (4.1), such that the following can be proved:

Proposition 4.1 *Suppose that $\hat{X}$ satisfies the assumptions of Theorem* 4.1 *while n, m fulfil* (4.1). *Then there is an integer r such that*

$$2m + 2 = (L - 1)(n + 1) + r, \quad 0 < r < n + 1, \tag{4.4}$$

and for

$$\tau := \epsilon c w^{-n-1} + \delta c w^{-r} + \epsilon \overline{c} w^{n+1} + \delta \overline{c} w^r, \quad \epsilon > 0,\ \delta > 0,\ c \in \mathbb{C}, \tag{4.5}$$

and for suitable choices of $D_t^\beta(\phi)(0)$ *with* $1 \le \beta \le \frac{1}{2}(L-2)$ *and* $D_t^\beta \phi(0) = 0$ *for* $\beta > \frac{1}{2}(L-2)$ *we obtain for* $L > 4$ *that*

$$E^{(L)}(0) = \mathrm{Re}\left\{\epsilon^{L-1}\int_{S^1} c^L \kappa \frac{dw}{w} + O(\epsilon^L)\right\} \quad \textit{where } \kappa := i^{L-1} p_L(n,m), \tag{4.6}$$

while $E^{(j)}(0) = 0$ *for* $1 \le j \le L-1$.

Thus we can achieve (4.3) for $k = L$ if $L > 4$ and $p_L(n,m) \neq 0$. If $p_L(n,m) = 0$ then (4.6) leads to nothing, and we do not see how (4.3) can be achieved. Yet by modifying our choice of τ and $D_t^\beta \phi(0)$ we obtain variations $\hat{Z}(t,\epsilon)$ of $\hat{X}$ depending on t and on $\epsilon > 0$ such that

$$E(t,\epsilon) := D(\hat{Z}(t,\epsilon))$$

satisfies the following

Proposition 4.2 *Under the preceding assumptions there are sequences* $\{t_\ell\}$ *and* $\{\epsilon_\ell\}$ *with* $0 < t_\ell \to 0$ *and* $0 < \epsilon_\ell \to 0$ *such that*

$$E(t_\ell, \epsilon_\ell) < E(0,0) = D(\hat{X}) \quad \textit{for } \ell \in \mathbb{N}. \tag{4.7}$$

Thus $\hat{X}$ *is not a weak relative minimizer of* D.

Finally, according to Theorem 2.5 in Chap. 2 and by virtue of the observation that (C_4) is equivalent to (T_4), we know that $\hat{X}$ cannot be a weak relative minimizer of D if n, m satisfy (4.1) with $L = 4$. In conjunction with Proposition 4.1 and 4.2 we arrive at

Theorem 4.3 *Suppose that* $\hat{X}$ *satisfies the assumptions of Theorem* 4.1 *while* n, m *fulfil condition* (4.1). *Then* $\hat{X}$ *cannot be a weak relative minimizer of* D.

Clearly, Theorem 4.1 is now a consequence of Theorems 4.2 and 4.3. Thus it remains to prove Theorem 4.2 and Propositions 4.1 and 4.2. We note that Theorem 4.2 is proved for $L = 3, 5$ and Theorem 4.3 for $L = 4$ (cf. Theorems 4.1, 4.2, 4.4 and 4.5). Thus *it suffices to consider* (4.1) *for* $L \ge 6$ *and* (4.2) *for* $L \ge 5$. We begin with

The first main case: L is an odd integer ≥ 5 satisfying (4.2)

Case I: $2m + 2 = L(n+1)$, L an odd integer ≥ 5. Here we choose

$$\tau = cw^{-n-1} + \bar{c}w^{n+1}, \quad c \in \mathbb{C}. \tag{4.8}$$

The, by now, standard computations yield

$$\begin{aligned} w\hat{Z}_{tw}(0)\tau = ic^2(&A_2 w^{-n} + \cdots + (2m-2n)A_{2m-2n+1}w^{2m-3n-1} + \cdots, \\ &(m-n)R_m w^{m-2n-1} + \cdots). \end{aligned}$$

Set $A_2 = \mu A_1$, $\mu \in \mathbb{C}$. Note that the number of terms that could possibly contain poles in the first two complex components are associated with coefficients $A_\ell, \ell \leq 2m - 2n$. This follows from the fact that

$$2m - 3n - 1 = 2(m+1) - 3(n+1) = (L-3)(n+1) \geq 0.$$

Consider the expression

$$f := w\hat{Z}_{tw}(0)\tau + w\hat{X}_w\phi_t(0). \tag{4.9}$$

We can make it holomorphic by eliminating the poles of $w\hat{Z}_{tw}(0)\tau$ step by step with an appropriate choice of $\phi_t(0)$ which has the form

$$\phi_t(0) := -i\mu c^2 w^{-2n-1} + i\,\overline{\mu}\,\overline{c}^2 w^{2n+1} + \text{pole terms of lower order.} \tag{4.10}$$

Consider the formulae (7.1)–(7.5) in Sect. 7.2. We have

$$E^{(L)}(0) = J_1 + J_2 + J_3,$$

and $J_3 = 0$ because of $\hat{X}_w \cdot \hat{X}_w = 0$ and $\hat{Z}_{tw}(0) \cdot \hat{X}_w = 0$. We continue to select $D_t^\beta \phi(0)$ in such a way that all integrands in J_1 are holomorphic, and so $J_1 = 0$. This is summarized in

Lemma 4.1 *By the pole-removal technique we can inductively choose $D_t^\beta\phi(0)$ for $\beta \leq \frac{1}{2}(L-3)$ such that g_ν is holomorphic for $\nu = 0, 1, \ldots, \frac{1}{2}(L-3)$. Then the holomorphic derivative $[D_t^\gamma \hat{Z}(0)]_w$ can be obtained in the form*

$$[D_t^\gamma \hat{Z}(0)]_w = \{ig_{\gamma-1}\}_w \quad \textit{for } \gamma = 1, 2, \ldots, \frac{1}{2}(L-1). \tag{4.11}$$

(*Note that this result is the analogue to Lemma* 3.1 *of Chap.* 3, *and it will be proved in a similar way.*)

Proof The pole-removal process can be carried on as long as the w-power attached to $A_{2m-2n+1}$ does not become negative. At the αth stage in defining $D_t^\alpha \hat{Z}(0)$ the powers are reduced by $\alpha(n+1)$; so we must check whether

$$2m - n + 1 - \alpha(n+1) \geq 0 \quad \text{for } \alpha \leq \frac{1}{2}(L-3).$$

In fact, this is even true for $\alpha \leq \frac{1}{2}(L+1)$ since

$$\begin{aligned}
2m - n + 1 - \left[\frac{1}{2}(L+1)\right](n+1) &= \frac{1}{2}\{4m - 2n + 2 - (L+1)(n+1)\} \\
&= \frac{1}{2}\{2(2m+2) - 2 - 2n - (L+1)(n+1)\} \\
&= \frac{1}{2}\{(2m+2) - 3(n+1)\} > 0.
\end{aligned}$$

From (4.10) it follows that

$$2w\hat{Z}_{tw}(0)\phi_t(0) = 2c^3(\mu^2 A_1 w^{-2n} + \cdots, \ldots).$$

Thus we have to choose

$$\phi_{tt}(0) := -2c^3\mu^2 w^{-3n-1} + \cdots \tag{4.12}$$

in order to remove the pole in (4.12).

We make four observations, which we call our "Fundamental Computational Principles":

1. The highest pole order of $w[D_t^2\hat{Z}(0)]_w\tau$ is at most n and thus strictly less than $2n+1$.
2. Inductively we see that the order of the leading pole term in $D_t^\beta\phi(0)$ needed to remove the poles in the first complex components of g_ν is just $(\beta+1)n+1$ with $\beta=\nu$; more precisely,

$$D_t^\beta\phi(0) := \text{const}\cdot c^{\beta+1}(-i)^\beta w^{-n(\beta+1)-1} + \cdots$$

   which is a consequence of the fact that

$$D_t^\beta\phi(0) = \beta!(-i)^\beta\tau(\tau_w + (n+1)\tau/w)^\beta + \cdots. \tag{4.13}$$

   To see this we focus on the term

$$w\hat{Z}_w(0)D_t^\beta\phi(0) \quad \text{for } \beta = \frac{1}{2}(L-3);$$

   the other terms yield the same result. Then the third complex component of g_ν has the leading term

$$\text{const}\, R_m w^{m+1}(-i)^\beta w^{-n(\beta+1)-1} = \text{const}\, R_m w^{m-n(\beta+1)}$$

   with $\beta=\nu$. But

$$\begin{aligned} m - n(\beta+1) &= (m+1) - (\beta+1)(n+1) + \beta \\ &= \frac{1}{2}\{2(m+1) - 2(\beta+1)(n+1)\} + \beta \\ &= \frac{1}{2}[L - 2(\beta+1)](n+1) + \beta \geq n+1+\beta > 0 \end{aligned}$$

   for $\beta \leq \frac{1}{2}(L-3)$.
3. For $\beta > \frac{1}{2}(L-3)$ we may assume $D_t^\beta\phi(0) = 0$.
4. In general, as we have seen, with one generator $1/z^{n+1}$ ($2m+2 = k(n+1)$, k odd), derivatives $D_t^\gamma\phi(0)$, $\gamma \geq 1$ do not affect the calculation of the Lth derivative; if we have two generators $1/w^{n+1}$, $1/w^r$ the $D_t^\gamma\phi(0)$, $\gamma \geq 2$ are inconsequential, and for three generators $1/w^{n+1}$, $1/w^r$, $1/w^s$ we may ignore the $D_t^\gamma\phi(0)$, $\gamma \geq 3$. This formula follows immediately from (4.13), since the leading term of $\tau_w + (n+1)\tau/w$ is zero. □

Thus we have found

$$E^{(L)}(0) = J_2.$$

Next we prove as in Chap. 3:

Lemma 4.2 *For the integrals $I_{\alpha\gamma\beta}$ defined by* (3.25) *of Chap.* 3 *we have*

$$I_{\alpha\gamma\beta} = 0 \quad \text{for } 1 \le \alpha,\ \gamma \le \frac{1}{2}(L-1),\ \beta \ge 1,\ \alpha+\beta+\gamma = L-1. \tag{4.14}$$

This implies

$$E^{(L)}(0) = \frac{2(L-1)!}{M!M!} \operatorname{Re} \int_{S^1} w[D_t^M \hat{Z}(0)]_w \cdot [D_t^M \hat{Z}(0)]_w \tau \, dw \quad \text{for } M := \frac{1}{2}(L-1). \tag{4.15}$$

Proof We simply count orders of zeros and poles. The lowest w-power of $[D_t^\alpha \hat{Z}(0)]_w$ and $[D_t^\gamma \hat{Z}(0)]_w$ is

$$m - \alpha(n+1) \quad \text{and} \quad m - \gamma(n+1) \quad \text{respectively.}$$

Thus the lowest w-power arising from the third complex components in

$$w[D_t^\alpha \hat{Z}(0)]_w \cdot [D_t^\gamma \hat{Z}(0)]_w D_t^\beta \phi(0), \quad \alpha+\beta+\gamma = L-1,\ 1 \le \alpha,\ \gamma \le \frac{1}{2}(L-1)$$

is by $(n+1)L = 2m+2$

$$\begin{aligned}
&(2m+1) - (\alpha+\gamma)(n+1) - (\beta+1)n - 1 \\
&\quad = (2m+2) - (\alpha+\beta+\gamma)(n+1) + (\beta - n - 1) \\
&\quad = \beta - 1 \ge 0 \quad \text{for } \beta \ge 1.
\end{aligned}$$

Thus there is no pole arising from the third component. What about the first two components? The "dangerous" contributions, by construction, are of the form

$$\text{const}\, A_j \cdot A_{2m-2n+1} w^{2m-n-\gamma(n+1)-(\beta+1)n} + \cdots,$$

or γ interchanged with α. Then

$$\begin{aligned}
&2m - n - \gamma(n+1) - (\beta+1)n \\
&\quad = 2m + 2 - (\gamma+\beta+1)(n+1) - n + (\beta+1) - 2 \\
&\quad = (2m+2) - (\gamma+\beta+2)(n+1) + \beta \\
&\quad = (2m+2) - (L - \alpha + 1)(n+1) + \beta \\
&\quad = (\alpha - 1)(n+1) + \beta \ge \beta \quad \text{since } \alpha \ge 1,
\end{aligned}$$

and similarly for α interchanged with γ. Thus there are no poles arising from the first two complex components. □

Lemma 4.3 *In* (4.15) *the contribution of the scalar product coming from the first two complex components is zero.*

Proof This contribution has as worst pole term an expression of the form

$$\text{const}\, A_j \cdot A_{2m-2n+1} w^{2m-n-[(L-1)/2](n+1)-(n+1)+1},$$

where j is some index between 1 and $2m-2n$. The power

$$2m - n - [(L-1)/2](n+1)$$

arises from the fact that, in the expressions $[D_t^\alpha \hat{Z}(0)]_w$, powers are reduced by $\alpha(n+1)$ from their initial expression $\hat{X}_w = (A_1 w^n + \cdots, R_m w^m + \cdots)$. However,

$$\begin{aligned} &2m - n - [(L-1)/2](n+1) - (n+1) + 1 \\ &\quad = \frac{1}{2}\{4m - 2n - (L-1)(n+1) - 2n\} \\ &\quad = \frac{1}{2}\{(2m+2) - (L-1)(n+1) + 2m - 4n - 2\} \\ &\quad = \frac{1}{2}\{(n+1) + 2m - 4n - 2\}. \end{aligned}$$

Since we assume $\frac{1}{2}(L-1) = M > 1$, we have $L > 3$ and therefore

$$\{(n+1) + 2m - 4n - 2\} = (2m+2) - 3(n+1) = (L-3)(n+1) > 0. \qquad \square$$

Lemma 4.4 *In* (4.15) *the only contribution of the scalar product coming from the product of the last complex components is the term*

$$\mu_0 R_m^2 c^L w^{-1} \quad \textit{with } \mu_0 := i^M \prod_{\ell=1}^{M} [(m+1) - \ell(n+1)]. \tag{4.16}$$

Proof First one realizes that the lowest w-power of

$$[D_t^\gamma \hat{Z}(0)]_w = \left\{ i \sum_{\alpha+\beta=\gamma-1} \frac{(\gamma-1)!}{\alpha!\beta!} w[D_t^\alpha \hat{Z}(0)]_w D_t^\beta \phi(0) \right\}_w$$

in the last complex component occurs strictly in the first term $w[D_t^{\gamma-1}\hat{Z}(0)]_w \tau$ since the pole associated with $D_t^\beta \phi(0)$ is of the order $(\beta+1)(n+1)+1$.

To prove (4.16) we start with $R_m w^{m+1}$, multiply by τ, which in the leading term is equivalent to dividing by w^{n+1}, and then differentiate obtaining $i(m-n)R_m w^{m-n-1}$. Continuing this process M times we obtain for the third complex component of $[D_t^M \hat{Z}(0)]_w$, $M = \frac{1}{2}(L-1)$, the expression

$$\mu_0 R_m c^M w^{m-M(n+1)} \quad \text{with } \mu_0 := i^M \prod_{\ell=1}^{M} [(m+1) - \ell(n+1)].$$

But

$$\begin{aligned} m - M(n+1) &= \frac{1}{2}\{2m - (L-1)(n+1)\} \\ &= \frac{1}{2}\{(2m+2) - (L-1)(n+1) - 2\} = \frac{1}{2}(n-1). \end{aligned}$$

Thus the third complex component of $[D_t^M \hat{Z}(0)]_w$ with $M = \frac{1}{2}(L-1)$ has the form

$$\mu_0 c^M R_m w^{(n-1)/2} + \text{higher order terms.}$$

From this we infer the assertion of Lemma 4.4. $\square$

By virtue of (4.15) and Lemmas 4.3 and 4.4 we arrive at

Proposition 4.3 *If $2m+2 = L(n+1)$, L odd and ≥ 5, then by the choice $\tau = cw^{-n-1} + \overline{c}w^{n+1}$, $c \in \mathbb{C}$, and by suitable choices of $D_t^\beta \phi(0)$ for $\beta \geq 1$ we obtain $E^{(j)}(0) = 0$ for $j = 1, \dots, L-1$ and*

$$E^{(L)}(0) = \frac{2(L-1)!}{M!M!} \operatorname{Re} \int_{S^1} \kappa c^L R_m^2 \frac{dw}{w}, \quad M := \frac{1}{2}(L-1), \tag{4.17}$$

with

$$\kappa := i^{L-1}(m-n)^2(m-2n-1)^2(m-3n-2)^2 \dots (m - Mn - (M-1))^2.$$

Since $m+1 = \frac{1}{2}L(n+1) = M(n+1) + \frac{1}{2}(n+1)$, we have

$$m - Mn - (M-1) = m + 1 - M(n+1) = \frac{1}{2}(n+1) > 0,$$

and therefore $\kappa \neq 0$. Hence

$$E^{(L)}(0) = \frac{2 \cdot (L-1)!}{M!M!} \operatorname{Re}\,[2\pi i \kappa c^L R_m^2] < 0 \tag{4.18}$$

for a suitable choice of c. Thus we have proved Theorem 4.2 in the case $2m+2 = L(n+1)$.

Remark 4.1 The beautiful formula (4.17) tells us that for the case $2m+2 = L(n+1)$ there is a Diophantine polynomial $p_L(x, y)$ such that $\kappa = i^{L-1} p_L(n, m)$, and that $p_L(n, m) \neq 0$. Note that p_L only depends on L, m and n and not on the specific minimal surface.

Remark 4.2 When we treated the case n odd, m even in the preceding section, we omitted the special case $n = 1$. We note that this case is included in Proposition 4.3, since then we have $n+1 = 2$ and therefore $2m+2 = L(n+1)$ with $L = m+1$.

Case II: L is odd and $2(m+1) = (L-1)(n+1) + r$ with $0 < r \leq n$. Here we choose

$$\tau := \epsilon c w^{-n-1} + c\delta w^{-r} + \epsilon \overline{c} w^{n+1} + \overline{c}\,\delta w^r \quad \text{with } \epsilon > 0,\ \delta > 0, \text{ and } c \in \mathbb{C}. \tag{4.19}$$

Then we prove the following result which is the analogue of Proposition 4.3:

Proposition 4.4 *If L is odd, $2m+2 = (L-1)(n+1) + r$, $0 < r \leq n$, and τ is chosen by (4.19), then by a suitable choices of $D_t^\beta \phi(0)$ for $\beta \geq 1$ we obtain $E^{(j)}(0) = 0$ for $j = 1, \dots, L-1$ and*

$$E^{(L)}(0) = \frac{2 \cdot (L-1)!}{M!M!} \epsilon^{L-1} \operatorname{Re} \int_{S^1} \delta c^L \kappa R_m^2 \frac{dw}{w} + O(\epsilon^L), \quad M := \frac{1}{2}(L-1), \tag{4.20}$$

where the constant κ depends only on L, r, m, n, $\kappa = i^{L-1} P_L(m, n\tau)$, the minimal surface polynomial.

To begin our discussion, again we have $E^{(L)}(0) = J_1 + J_2 + J_3$, and $J_3 = 0$ since $\hat{X}$ is a minimal surface. Furthermore we can show the analogue of Lemma 4.1: *The* $g_{\gamma-1}$ *are holomorphic for* $\gamma = 1, 2, \ldots, \frac{1}{2}(L-1)$ if we choose ϕ_t meromorphic on B, real on S^1, and such that

$$\phi_t(0) := -i\mu c^2\epsilon^2 w^{-2n-1} - i\delta^2 c^2 \epsilon(n+1-r)w^{-n-1-r} + \cdots \quad (\text{if } A_2 = \mu A_1) \tag{4.21}$$

and then inductively $D_t^\beta \phi(0)$ for $\beta > 1$, following the construction of Lemma 4.1:

$$\begin{aligned} D_t^\beta \phi(0) &= \text{const} \cdot (\epsilon c)^{\beta+1}(-i)^\beta w^{-(\beta+1)n-1} \\ &\quad + \text{const} \cdot \epsilon^\beta c^{\beta+1}(-i)^\beta w^{-\beta n - r - 1} + \cdots \quad \text{for } \beta \le \frac{1}{2}(L-3) \end{aligned} \tag{4.22}$$

and

$$D_t^\beta \phi(0) = 0 \quad \text{for } \beta \ge \frac{1}{2}(L-1). \tag{4.23}$$

This implies $J_1 = 0$, and we are left with

$$E^{(L)}(0) = J_2. \tag{4.24}$$

In order to reduce J_2 to two terms we can use the following

Lemma 4.5 *After the pole-removal procedure we obtain*

$$\begin{aligned} &w[D_t^\alpha \hat{Z}(0)]_w \cdot [D_t^\gamma \hat{Z}(0)]_w D_t^\beta \phi(0) \\ &\quad = \epsilon^{L-1}\{Cw^{\beta-2} + \cdots\} + O(\epsilon^L) + P(\epsilon, w) \end{aligned} \tag{4.25}$$

for $\alpha + \beta + \gamma + 1 = L$, $0 \le \alpha, \gamma \le \frac{1}{2}(L-1)$, $1 \le \beta \le \frac{1}{2}(L-3)$, *where the remainder term* $P(\epsilon, w)$ *involves terms of lower order in* ϵ, *but is holomorphic in* w, *i.e.* $\int_{S^1} P(\epsilon, w)\, dw = 0$, *and* $C \in \mathbb{C}$.

Proof (i) First we look at the contribution from the third components to the left-hand side of (4.25). At each stage of the pole-removal process we are essentially successively dividing by w^{n+1} and w^r (since the multiplication by w balances differentiation in considering the order of the resulting exponents) and removing poles in the first two complex components. If we wish to look at terms of order ϵ^{L-1} we need to consider one contribution from $\delta c w^{-r}$ in the definition of $\tau = \phi(0)$ or const w^{-n-1-r} in the definition of $\phi_t(0)$. A greater contribution from $\delta c w^{-r}$ would result in a larger exponent in w but a lower exponent in ϵ.

If this contribution comes from the $D_t^\beta \phi(0)$-term, then the leading power of the contribution from the product of the last complex components of (4.25) will be

$$\begin{aligned} &1 + 2m - (\alpha+\gamma)(n+1) - \beta n - r - 1 \\ &\quad = (2+2m) - (\alpha+\beta+\gamma)(n+1) - r + \beta - 2 = \beta - 2. \end{aligned} \tag{4.26}$$

If, however, the contribution arises from $[D_t^\alpha \hat{Z}(0)]_w$, $[D_t^\gamma \hat{Z}(0)]_w$, then the leading power from the product of the two last components of (4.25) will be

$$\begin{aligned} &1+2m-(\alpha+\gamma-1)(n+1)-r-(\beta+1)n-1 \\ &\quad=(2+2m)-(\alpha+\beta+\gamma)(n+1)-r+\beta-1=\beta-1. \end{aligned} \tag{4.27}$$

Note that $\beta-2\geq 0$ for $\beta\geq 2$ and $\beta-1\geq 0$ for $\beta\geq 1$.

The first two complex components of $[D_t^\alpha \hat{Z}(0)]_w$ are $(F_\alpha+\cdots)\in\mathbb{C}^2$. Thus, if the worst contribution of $\delta c w^{-r}$ arises from the $D_t^\beta\phi(0)$-term, the leading term of the contributions of the first two components to (4.25) is of the form

$$\begin{aligned} &\epsilon^{\alpha+\beta+\gamma}[F_\alpha\cdot A_{2m-2n+1}]w^{1+2m-n-\gamma(n+1)-\beta n-r-1} \\ &\quad+\epsilon^{\alpha+\beta+\gamma}[F_\gamma\cdot A_{2m-2n+1}]w^{1+2m-n-\alpha(n+1)-\beta n-r-1}. \end{aligned}$$

But, for $\alpha\geq 1$,

$$\begin{aligned} &1+2m-n-\gamma(n+1)-\beta n-r-1 \\ &\quad=2+2m-(\gamma+\beta+1)(n+1)+\beta-r-1 \\ &\quad=2m+2-(L-\alpha)(n+1)-r+(\beta-1) \\ &\quad\geq(2m+2)-(L-1)(n+1)-r+(\beta-1)=\beta-1\geq 0 \quad \text{if } \beta\geq 1, \end{aligned}$$

and similarly

$$1+2m-n-\alpha(n+1)-\beta n-r-1\geq\beta-1\geq 0 \quad \text{if } \gamma\geq 1 \text{ and } \beta\geq 1.$$

Furthermore, if $\gamma=0$, then $\beta=L-1-\alpha\geq\frac{1}{2}(L-1)$ whence $D_t^\beta\phi(0)=0$, and so the left-hand side of (4.25) is zero, i.e. (4.25) holds trivially, and similarly for $\alpha=0$.

If the worst contribution arises either from $[D_t^\alpha\hat{Z}(0)]_w$ or from $[D_t^\gamma\hat{Z}(0)]_w$, then the leading power will be the minimum of the two numbers

$$\begin{aligned} &1+2m-n-(\gamma-1)(n+1)-r-(\beta+1)n-1, \\ &1+2m-n-(\alpha-1)(n+1)-r-(\beta+1)n-1. \end{aligned}$$

But

$$\begin{aligned} &1+2m-n-(\gamma-1)(n+1)-(\beta+1)n-r-1 \\ &\quad=2+2m-(\gamma+\beta+1)(n+1)-r+\beta \\ &\quad=2+2m-(L-\alpha)(n+1)-r+\beta \\ &\quad\geq(2m+2)-(L-1)(n+1)-r+\beta=\beta\geq 0, \end{aligned}$$

and the same lower bound holds for the other term. Therefore, no pole comes from the first two complex components, and Lemma 4.5 is proved. □

An immediate consequence of Lemma 4.5 is, by Cauchy's theorem:

Lemma 4.6 *Under the hypotheses of Lemma* 4.5 *and the additional assumption* $\beta\geq 2$ *we obtain*

$$\int_{S^1} w[D_t^\alpha\hat{Z}(0)]_w\cdot[D_t^\gamma\hat{Z}(0)]_w D_t^\beta\phi(0)\,dw=O(\epsilon^L). \tag{4.28}$$

Proposition 4.5 *Under the assumptions of Proposition* 4.4 *we obtain* $E^{(j)}(0)=0$ *for* $j=1,\dots,L-1$ *and with* $M=\frac{1}{2}(L-1)$:

$$E^{(L)}(0)=\frac{2(L-1)!}{M!M!}\epsilon^{L-1}\operatorname{Re}\left\{\int_{S^1}w[D_t^M\hat{Z}(0)]_w\cdot[D_t^M\hat{Z}(0)]_w\tau\right\}dw$$
$$+(L-1)\operatorname{Re}\int_{S^1}w[D_t^M\hat{Z}(0)]_w\cdot[D_t^{M-1}\hat{Z}(0)]_w\phi_t(0)\,dw+O(\epsilon^L). \tag{4.29}$$

Proof According to Lemma 4.5, the left-hand side of (4.25) has poles only for $\beta=0$ and $\beta=1$. Inspecting formula (3.4) in Chap. 3, this leaves just two terms for J_2, and by virtue of (4.24) we arrive at (4.29). □

Equation (4.29) is the fundamental formula, or normal form for odd order derivatives of Dirichlet's energy. To show that we can actually calculate (4.29) we need the following critical lemma:

Lemma 4.7 *If* L *is either odd or even and* r *is defined by either* (4.1) *or* (4.2), *then taking our generator as* $\tau=c\epsilon/w^{n+1}+\delta c/w^r+\overline{c}+w^{n+1}+\overline{\delta}\overline{c}w^r$, *we can calculate the leading terms (in* ϵ *and* w*) of the* **last complex** *component of* $w[D_t^k\hat{Z}(0)]_w$ *for either* $k\le\frac{L-1}{2}$ *(*L *odd), or* $k=\frac{L}{2}-1$ *(*L *even) by*

$$\begin{aligned}&c^ki^k\epsilon^k[m+1-(n+1)][m+1-2(n+1)]\cdot\dots\\&\cdot[(m+1-k(n+1)]R_mw^{m+1-k(n+1)}\\&+\delta c^ki^k\epsilon^{k-1}k[m+1-(n+1)]\cdot\dots\cdot[m+1-(k-1)(n+1)]\\&\cdot[m+1-(k-1)(n+1)-r]R_mw^{m+1-(k-1)(n+1)-r}.\end{aligned} \tag{4.30}$$

Proof By induction. The statement is clearly true for $k=1$. Assume it is true for k, and let us show that it holds for $k+1$.

By definition:

$$\begin{aligned}&w[D_t^k\hat{Z}(0)]_w\phi\\&=c^{k+1}i^k\epsilon^{k+1}[(m+1)-(n+1)]\cdot\dots\\&\cdot[m+1-k(n+1)]R_mw^{m+1-(k+1)(n+1)} &(4.31)\\&+\delta c^{k+1}i^k\epsilon^kk[m+1-(n+1)]\cdot\dots\cdot[m+1-(k-1)(n+1)]\\&\cdot[m+1-(k-1)(n+1)-r]R_mw^{m+1-k(n+1)-r} &(4.32)\\&+\delta c^{k+1}i^k\epsilon^k[(m+1)-(n+1)]\cdot\dots\\&\cdot[m+1-k(n+1)]R_mw^{m+1-k(n+1)-r} &(4.33)\\&+\delta^2c^{k+1}i^k\epsilon^{k-1}k[m+1-(n+1)]\dots[m+1-(k-1)(n+1)]\\&\cdot[m+1-(k-1)(n+1)-r]R_mw^{m+1-(k-1)(n+1)-2r} &(4.34)\end{aligned}$$

using the fact that all derivatives $w[D_t^k \hat{Z}(0)]_w$ depend only on ϕ and ϕ_t (and not on higher derivatives in t of ϕ) we get (using $\phi_t = -\delta c^2 \epsilon i(n+1-r)/w^{n+1+r}$) that

$$kw[D_t^{k-1}\hat{Z}(0)]_w \phi_t \tag{4.35}$$

$$= -\delta c^{k+1} i^k \epsilon^k k[m+1-(n+1)] \cdot \ldots$$

$$\cdot [m+1-(k-1)(n+1)](n+1-r) R_m w^{m+1-k(n+1)-r} \tag{4.36}$$

$$- \delta^2 c^{k+1} i^k \epsilon^{k-1} k[m+1-(n+1)] \cdot \ldots$$

$$\cdot [m+1-(k-2)(n+1)-r](n+1-r) R_m w^{m+1(k-1)(n+1)-2r}. \tag{4.37}$$

Now (4.32) + (4.36) equals (since $[m+1-(k-1)(n+1)-r] - [(n+1)-r] = [m+1-k(n+1)]$)

$$\delta c^{k+1} i^k \epsilon^k k[m+1-(n+1)] \ldots [m+1-k(n+1)] R_m w^{m+1-k(n+1)-r} \tag{4.38}$$

and (4.32) + (4.36) + (4.33) equals

$$\delta c^{k+1} i^k \epsilon^k (k+1)[(m+1-(n+1)] \ldots [(m+1-k(n+1)] R_m w^{m+1-k(n+1)-r}. \tag{4.39}$$

Since (4.34) and (4.37) are lower order terms we see that the leading terms of the last complex component

$$w[D_t^{k+1}\hat{Z}(0)]_w = [w[D_t^k \hat{Z}(0)]_w \phi + w[D_t^{k-1}\hat{Z}]_w \phi_t]_w \tag{4.40}$$

= (4.31) + (4.39) which proves the lemma. □

We can now prove a theorem stating explicitly the Lth derivative of Dirichlet's energy if L is odd.

Proposition 4.6 *If $L \geq 3$ is odd, $r \leq n$, the minimal surface polynomial $P_L(m,n,\tau)$ (we include τ to indicate the dependence of the minimal surface polynomial on the generator τ. (cf. (4.20))) does not depend on r and is given by*

$$\begin{aligned} P_L(m,n,r) &=: P_L(m,n) \\ &= \frac{2L!}{\left\{\left(\frac{L-1}{2}\right)!\right\}^2}[m-n]^2[1+m-2(n+1)]^2 \cdot \ldots \\ &\quad \cdot \left[1+m-\left(\frac{L-1}{2}\right)(n+1)\right]^2. \end{aligned} \tag{4.41}$$

Remark Compare this with Wienholtz's formula (Theorem 2.1) of Chap. 2 and formula (2.58) of Chap. 2. We thus have beautiful formulae for the Lth derivative (L odd) of Dirichlet's energy.

Theorem 4.4 *If $L \geq 3$ is odd, $r \leq n$, the Lth derivative of Dirichlet's energy is given by*

$$E^{(L)}(0) = \epsilon^{L-1} \operatorname{Re} \int \delta c^L \cdot i^{L-1} P_L(m,n) R_m^2 \frac{dw}{w} + O(\epsilon^L). \tag{4.42}$$

Proof The proofs of Proposition 4.6 and formula (4.42) follows from Lemma 4.6, since we have an explicit formula for the last complex component of $[D_t^k \hat{Z}(0)]_w$. By Lemma 4.7, we have, setting $k = \frac{L-1}{2}$ and $k = \frac{L-3}{2}$ formulae for $w[D_t^{\frac{L-1}{2}} \hat{Z}(0)]_w$ and $w[D_t^{\frac{L-3}{2}} \hat{Z}(0)]_w$.

Using these formulae we obtain that the last complex component of

$$\begin{aligned}
&w^2[D_t^{\frac{L-1}{2}} \hat{Z}(0)]_w \cdot [D_t^{\frac{L-1}{2}} \hat{Z}(0)]_w \\
&\quad = c^{L-1}\epsilon^{L-1}i^{L-1}[m-n]^2[1+m-2(n+1)]^2 \cdot \ldots \\
&\qquad \cdot \left[1+m-\left(\frac{L-1}{2}\right)(n+1)\right]^2 R_m w^{2m+2-(L-1)(n+1)} \\
&\qquad + \delta(L-1)c^{L-1}\epsilon^{L-2}i^{L-1}[m-n]^2 \cdot \ldots \\
&\qquad \cdot \left[1+m-\left(\frac{L-3}{2}\right)(n+1)\right]^2 \left[1+m-\left(\frac{L-1}{2}\right)(n+1)\right] \\
&\qquad \cdot \left[1+m-\left(\frac{L-3}{2}\right)(n+1)-r\right] \\
&\qquad \cdot R_m w^{2m+2-(L-2)(n+1)-r} + \text{terms of lower order in } w.
\end{aligned} \tag{4.43}$$

Thus the last complex component of

$$\begin{aligned}
&w^2[D_t^{\frac{L-1}{2}} \hat{Z}(0)]_w \cdot [D_t^{\frac{L-1}{2}} \hat{Z}(0)]_w \phi \\
&\quad = c^L i^{L-1}\epsilon^L[m-n]^2 \cdot \ldots \cdot \left[1+m-\left(\frac{L-1}{2}\right)(n+1)\right]^2 \\
&\qquad \cdot R_m w^{2m+2-L(n+1)}
\end{aligned} \tag{4.44}$$

$$\begin{aligned}
&\qquad + \delta c^L i^{L-1}\epsilon^{L-1}[m-n]^2 \cdot \ldots \cdot \left[1+m-\left(\frac{L-1}{2}\right)(n+1)\right]^2 \\
&\qquad \cdot R_m w^{2m+2-(L-1)(n+1)-r}
\end{aligned} \tag{4.45}$$

$$\begin{aligned}
&\qquad + \delta(L-1)c^L i^{L-1}\epsilon^{L-1}[m-n]^2 \cdot \ldots \cdot \left[1+m-\left(\frac{L-3}{2}\right)(n+1)\right]^2 \\
&\qquad \cdot \left[1+m-\left(\frac{L-1}{2}\right)(n+1)\right] \cdot \left[1+m-\left(\frac{L-3}{2}\right)(n+1)-r\right] \\
&\qquad \cdot R_m w^{2m+2-(L-1)(n+1)-r} + \text{terms of lower order in } w.
\end{aligned} \tag{4.46}$$

Moreover the LCC (last complex component) of

$$\begin{aligned}
&w^2[D_t^{\frac{L-1}{2}} \hat{Z}(0)]_w \cdot [D_t^{\frac{L-3}{2}} \hat{Z}(0)]_w \\
&\quad = \delta c^{L-2}i^{L-2}\epsilon^{L-2}[m-n]^2 \cdot \ldots \cdot \left[1+m-\left(\frac{L-3}{2}\right)(n+1)\right]^2 \\
&\qquad \cdot \left[1+m-\left(\frac{L-1}{2}\right)(n+1)\right] R_m w^{2m+2-(L-2)(n+1)} \\
&\qquad + \text{terms of higher order in } w
\end{aligned}$$

yielding that the LCC of

$$
\begin{aligned}
&(L-1)w^2[D_t^{\frac{L-1}{2}}\hat{Z}(0)]_w \cdot [D_t^{\frac{L-3}{2}}\hat{Z}(0)]_w\phi_t(0) \\
&= -\delta(L-1)c^L i^{L-1}\epsilon^{L-1}[m-n]^2 \cdot \ldots \cdot \left[i+m-\left(\frac{L-3}{2}\right)(n+1)\right]^2 \\
&\cdot \left[1+m-\left(\frac{L-1}{2}\right)(n+1)\right](n+1-r) \\
&\cdot R_m w^{2m+2-(L-1)(n+1)-r} + \text{terms of higher order in } w.
\end{aligned} \tag{4.47}
$$

Noting that

$$
\left[1+m-\left(\frac{L-3}{2}\right)(n+1)-r\right]-(n+1-r)=\left[1+m-\left(\frac{L-1}{2}\right)(n+1)\right]
$$

we see that (4.46) + (4.47) equals

$$
\begin{aligned}
&\delta(L-1)c^L i^{L-1}\epsilon^{L-1}[m-n]^2 \cdot \ldots \cdot \left[1+m-\left(\frac{L-1}{2}\right)(n+1)\right] \\
&\cdot R_m w^{2m+2-(L-1)(n+1)-r} + \text{terms of higher order in } w
\end{aligned} \tag{4.48}
$$

and adding this to (4.45) yields that the LCC of

$$
w[D_t^{\frac{L-1}{2}}\hat{Z}(0)]_w \cdot \left\{[D_t^{\frac{L-1}{2}}\hat{Z}(0)]_w\phi + (L-1)[D_t^{\frac{L-3}{2}}\hat{Z}(0)]_w\phi_t(0)\right\}
$$

equals

$$
\delta L c^L \epsilon^{L-1} i^{L-1}[m-n]^2 \cdot \ldots \cdot \left[1+m-\left(\frac{L-1}{2}\right)(n+1)\right]^2 R_m/w + \cdots
$$

yielding Proposition 4.6 and Theorem 4.4. □

For $L > 5$, the same argument as in Lemma 4.3 shows that the Lth derivative does not depend on the first complex components of $\hat{X}_w$.

The second main case: L is an even integer ≥ 6

As we have stated at the beginning we must prove Propositions 4.1 and 4.2 in order to verify Theorem 4.3, which then completes the proof of Theorem 4.1. Recall that we now have

$$
2m+2=(L-1)(n+1)+r, \quad 1 \leq r \leq n, \ L \text{ even}. \tag{4.49}
$$

We will use the following modification of formulae (3.1)–(3.5) of Chap. 3, employing the same definitions for g_ν, $T^{\alpha,\beta}$, h_σ, and $\psi(M,\alpha)$ as before

$$
E^{(L)}(0) = J_1 + J_2 + J_3 + J_4 + J_5 \tag{4.50}
$$

with

$$J_1 := 4\,\mathrm{Re} \int_{S^1} w[D_t^{L-1}\hat{Z}(0)]_w \cdot \hat{X}_w \tau \, dw;$$

$$J_2 := \sum_{M=s+1}^{L-2} \frac{4(L-1)!}{M!(L-M-1)!} \mathrm{Re} \int_{S^1} [D_t^M \hat{Z}(0)]_w \cdot g_{L-M-1} \, dw, \quad s := L/2;$$

$$J_3 := \frac{4(L-1)!}{s!(s-1)!} \mathrm{Re} \int_{S^1} [D_t^s \hat{Z}(0)]_w \cdot g_{s-1} \, dw$$

$$+ \frac{2(L-1)!}{\sigma!\sigma!} \mathrm{Re} \int_{S^1} [D_t^\sigma \hat{Z}(0)]_w \cdot h_\sigma \, dw, \quad \sigma = s-1 = L/2 - 1;$$

$$h_\sigma := \sum_{\alpha=0}^{\sigma} c_{\alpha\beta}^\sigma \psi(\sigma,\alpha) w [D_t^\alpha \hat{Z}(0)]_w D_t^\beta \phi(0),$$

$$\beta = (L-1) - (s-1) - \alpha = L - s - \alpha;$$

$$g_\nu := \sum_{\alpha+\beta=\nu} c_{\alpha\beta}^\nu T^{\alpha,\beta} \quad \text{with } c_{\alpha\beta}^\nu := \frac{\nu!}{\alpha!\beta!}, \ \alpha+\beta+\nu = L-1;$$

$$T^{\alpha,\beta} := w(D_t^\alpha \hat{Z}(0))_w D_t^\beta \phi(0);$$

$$J_4 := \sum_{M=2}^{s-2} \frac{2(L-1)!}{M!M!} \mathrm{Re} \int_{S^1} [D_t^M \hat{Z}(0)]_w \cdot h_M \, dw$$

$$+ \frac{2(L-1)!}{(L-3)!} \mathrm{Re} \int_{S^1} w \hat{Z}_{tw}(0) \cdot \hat{Z}_{tw}(0) D_t^{L-3} \phi(0) \, dw, \quad \alpha+\beta+M = L-1;$$

$$J_5 := 4(L-1)\,\mathrm{Re} \int_{S^1} w \hat{Z}_{tw}(0) \cdot \hat{X}_w D_t^{L-2} \phi(0) \, dw$$

$$+ 2\,\mathrm{Re} \int_{S^1} w \hat{X}_w \cdot \hat{X}_w D_t^{L-1} \phi(0) \, dw.$$

The, by now, standard reasoning yields $J_5 = 0$, and $J_1 = 0$ if $\tau = \phi(0)$ is the generator of an inner forced Jacobi field. We choose

$$\tau := c\epsilon w^{-n-1} + \delta c w^{-r} + \overline{c}\epsilon w^{n+1} + \delta \overline{c} w^r, \quad \epsilon > 0, \ \delta > 0, \tag{4.51}$$

$c \in \mathbb{C}$, and $\phi_t(0)$ meromorphic, real on S^1, and

$$\phi_t(0) := -i\mu c^2 \epsilon^2 w^{-2n-1} - i\delta c^2 \epsilon (n+1-r) w^{-n-1-r} + \cdots, \tag{4.52}$$

if $A_2 = \mu A_1$, $\mu \in \mathbb{C}$. Then the pole-removal process leads to the following analogue of Proposition 4.4 which is just Proposition 4.1:

Assertion Under the assumption (4.49) the construction yields $E^{(j)}(0) = 0$ for $1 \le j \le L-1$ and

$$E^{(L)}(0) = \epsilon^{L-1} \mathrm{Re} \int_{S^1} i^{L-1} c^L \kappa R_m^2 \frac{dw}{w} + O(\epsilon^L) \tag{4.53}$$

where the constant κ depends only on L, r, n, m.

Although this is more or less straightforward from what we have already done in case II, a few remarks are appropriate. With our construction, $J_2 = 0$ since the integrands are holomorphic. The pole-removal is achieved by suitable choices of $D_t^\beta \phi(0)$ for $1 \le \beta \le L/2 - 1$, while we can set $D_t^\beta \phi(0) = 0$ for $\beta > L/2 - 1$. In J_4, the fact that $\beta \ge 3$, together with a minor variant of Lemma 3.1 in Chap. 3 show that $J_4 = O(\epsilon^L)$. This leaves the two terms of J_3.

Thus, we arrive at a formula for the Lth derivative, L even:

Proposition 4.7 *Taking generators as above, if L is even, the Lth derivative is given by*

$$E^{(L)}(0) = \frac{4(L-1)!}{M!(M-1)!}\epsilon^{L-1} \operatorname{Re} \int w[D_t^M \hat{Z}(0)]_w \cdot \{[D_t^{M-1}\hat{Z}(0)]_w \phi + (M-1) D_t^{M-2}\hat{Z}(0)\phi_t(0)\}\, dw \tag{4.54}$$

$$+ \frac{2(L-1)!}{(M-1)!(M-1)!}\epsilon^{L-1} \operatorname{Re} \int w[D_t^{M-1}\hat{Z}(0)]_w \cdot [D_t^{M-1}\hat{Z}(0)]_w \phi_t(0) \tag{4.55}$$

where $M = L/2 + O(\epsilon^L)$.

The following lemma, (the analogue of Lemma 2.5 in Chap. 2) allows us to compute (4.57).

Lemma 4.8 *Suppose that $f(w)$ is a meromorphic function in B which is of the form*

$$f(w) = \epsilon^s h(w) + \epsilon^{s-1} g(w) \tag{4.56}$$

with

$$h(w) = \sum_{j=1}^{N} \frac{a_j}{w^j} \quad \text{and} \quad g(w) = \sum_{k=N}^{\infty} b_k w^k, \quad N \in \mathbb{N}. \tag{4.57}$$

Then

$$\int_{S^1} \{2H[\operatorname{Re}\, if]\}_w \cdot f\, dw = -2\pi \epsilon^{2s-1} N a_N b_N + O(\epsilon^{2s}). \tag{4.58}$$

Proof By Chap. 2, Lemma 2.5, we have

$$\{2H[\operatorname{Re}(a_j w^{-j})]\}_w = j\bar{a}_j w^{j-1}.$$

This implies

$$\{2H[\operatorname{Re}(if)]\}_w = \epsilon^s \sum_{j=1}^{N} j(-i)\bar{a}_j w^{j-1} + \epsilon^{s-1} \sum_{k=N}^{\infty} k i b_k w^{k-1}.$$

Multiplication with f yields

$$\{2H[\operatorname{Re}(if)]\}_w \cdot f = \epsilon^{2s-1} i\, N\, a_N b_N w^{-1} + \epsilon^{2s} h(w) + k(w, \epsilon)$$

where $h(w)$ is meromorphic, and $k(w, \epsilon)$ is holomorphic. Integration over S^1 leads to (4.58). □

This says that, in evaluating (4.54), we take N times the product of the pole and zero terms of the highest and lowest orders respectively. We now state a version of Lemma 4.7 which unfortunately complicates the branch point story:

Lemma 4.9 *If $k = L/2$, the δ-linear term of $w[D_t^k \hat{Z}(0)]_w$ is given by*

$$\delta c^k i^k \epsilon^{k-1} k[m-n] \cdot \ldots \cdot [m+1-(k-1)(n+1)] \\ \cdot [(m+1-(k-1)(n+1)-r] R_m w^{m+1-(k-1)(n+1)-r}. \tag{4.59}$$

Proof Exactly as in Lemma 4.7. □

This now immediately implies (together with Lemma 4.8) that the ϵ^{L-1} term of (4.54) and (4.55) is zero; i.e. $k = 0$. This result appears as rather surprising to the author, even though we saw that this is true for the fourth derivative. Hence, we must either consider another generator, e.g. $\tau = c\epsilon/w^{n+1} + \delta c/w^s + \cdots$, $s = (n+1+r)/2$.

In this case we calculate the δ^2 terms of the Lth derivative or we could consider the $(L+1)^{\text{st}}$ derivative with our original generator $\tau = c\epsilon/w^{n+1} + \delta c/w^r$. This last possibility can be carried out but it is, in fact quite technical and so we leave this for the appendix. The simplest route is to change generators and show that, with this change, the Lth derivative can be made negative while all lower order derivatives vanish. We begin with the fourth derivative.

From Proposition 2.3 of Chap. 2 we have

$$\frac{d^4 E}{dt^4}(0) = 12\,\mathrm{Re} \int_{S^1} \hat{Z}_{ttw}(0) \cdot [w \hat{Z}_{tw}(0)\tau + w \hat{X}_w \phi_t(0)]\, dw \\ - 12\,\mathrm{Re} \int_{S^1} w^3 \hat{X}_{ww} \cdot \hat{X}_{ww} \tau^2 \phi_t(0)\, dw. \tag{4.60}$$

Take the generator $\tau = c\epsilon/w^{n+1} + \delta c/w^s + \cdots + s = (n+1+r)/2$. Then $2m+2 = 2(n+1) + 2s$ and

$$\phi_t = -\delta i c^2 \epsilon (n+1-s) R_m w^{-(n+1+s)} - \delta^2 c^2 i (n+1-s) R_m w^{-2s} + \cdots. \tag{4.61}$$

We now evaluate the δ^2-terms of the fourth derivative. Now the δ^2-term of the last complex component of $w[\hat{Z}_{tt}(0)]_w$ is given by

$$i\epsilon^2 (m-n) R_m w^{1+m-2(n+1)} \tag{4.62}$$

(there is no δ-linear pole).

Using Lemma 4.8, the first term of (4.60) equals

$$-12\,\mathrm{Re} \int i\epsilon^2 c^4 \delta^2 (m-n)^2 (1+m-2s) R_m^2 / w\, dw + \cdots \tag{4.63}$$

whereas the second term of (4.60) equals

$$-12\,\mathrm{Re}\int (m-n)^2\tau^2\phi_t R_m^2 w^{2m+1}\,dw+\cdots$$
$$=+36\,\mathrm{Re}\int ic^4\epsilon^2\delta^2(m-n)^2(n+1-s)R_m^2/w\,dw+\cdots. \qquad (4.64)$$

Noting that $1+m-2s=n+1-s$, we obtain

$$\frac{d^4E}{dt^4}(0)=+24\,\mathrm{Re}\int ic^4\epsilon^2\delta^2(m-n)^2(n+1-s)R_m^2/w\,dw+O(\epsilon^3) \qquad (4.65)$$

which can be made negative for an appropriate choice of c.

This completes the discussion of the fourth derivative with the new generators.

The disadvantage of these generators, in general, is that we must calculate both with $\phi_t(0)$ and $\phi_{tt}(0)$, and here our results (as the reader may verify) are independent of $D_t^\beta\phi(0)$, $\beta>2$. Now using the same reasoning as in Proposition 4.5, we obtain the following normal form result.

Proposition 4.8 (Normal form) *Suppose L is even, then $n+1+r$ is even ($2m+2=(L-2)(n+1)+(n+1+r)$). Let $s=(n+1+r)/2$ and consider the generator $\tau=c\epsilon/w^{n+1}+\delta/w^s+\overline{c}\epsilon w^{n+1}+\overline{\delta}w^s$. Then all the derivatives of E of order lower than L vanish and the Lth derivative has the form:*

$$\begin{aligned}E^L(0)=&\frac{4(L-1)!}{(L/2)!(L/2-1)!}\mathrm{Re}\int w[D_t^{L/2}\hat{Z}]_w\\&\cdot\{[D_t^{L/2-1}\hat{Z}(0)]_w\tau+(L/2-1)[D_t^{L/2-2}\hat{Z}(0)]\phi_t(0)\\&+\frac{(L/2-1)(L/2-2)}{2}[D_t^{L/2-3}\hat{Z}(0)]_w\phi_{tt}(0)\}\,dw\\&+\frac{2(L-1)!}{(L/2-1)!(L/2-1)!}\mathrm{Re}\int w[D_t^{L/2-1}\hat{Z}(0)]_w\\&\cdot\{[D_t^{L/2-1}\hat{Z}(0)]\phi_t(0)+(L/2-1)[D_t^{L/2-2}\hat{Z}(0)]_w\phi_{tt}(0)\}\,dw. \qquad (4.66)\end{aligned}$$

This result does **not** *depend on the first complex components.*

Again, we obtain

$$E^L(0)=\mathrm{Re}\left\{\epsilon^{L-2}\int_{S^1}c^L\delta^2kR_m^2\frac{dw}{w}\right\}+O(\epsilon^{L-1}) \qquad (4.67)$$

where $k=i^{L-1}P_L(m,n,\tau)$, $P_L(m,n,\tau)$ the minimal surface polynomial (including τ here to demonstrate dependence on the choice of generator).

We now have:

Proposition 4.9 *Suppose $L>4$ is even. Then $n+1+r$ is also even and $s=(n+1+r)/2$ is an integer. Let $\tau:=c\epsilon/w^{n+1}+c\delta/w^s+\tau\in w^{n-1}+\overline{c}\overline{\delta}w^s$. Then*

$$\begin{aligned}P_L(m,n,\tau)=&\frac{-L!}{(L/2-1)!(L/2-1)!}[m-n]^2\cdot\ldots\\&\cdot[1+m-(L/2-1)(n+1)]^2(n+1-s). \qquad (4.68)\end{aligned}$$

As a direct consequence we obtain the analogue of Theorem 4.2, namely

Theorem 4.5 *Suppose that $\hat{X}$ satisfies the assumption of Theorem* 4.1, *while n, m satisfy condition* (4.1). *Then with the choice of generator as in Proposition* 4.9, *we can achieve* (4.3) *for L even.*

Noting Theorem 4.4, we see that we have proved Theorem 4.1, which was our main goal.

Remark If $w = 0$ is exceptional, $r = (n+1)$ and $s = n+1$ and so formula (4.68) shows that the leading term of the Lth (even) derivative is zero.

We are now ready to proceed with the proof of Proposition 4.9.

Lemma 4.10 *Suppose $k \leq L/2$. Then the δ^2 term of the third complex component of $w[D_t^k \hat{Z}(0)]_w$ is given by*

$$i^k \frac{(k)(k-1)}{2} \epsilon^{k-2} c^k \delta^2 \rho_k R_m w^{1+m-(k-2)(n+1)-2s} + \cdots \tag{4.69}$$

where

$$\rho_k = [m-n] \ldots [1+m-(k-1)(n+1)][1+m-(k-2)(n+1)-2s]. \tag{4.70}$$

Proof (by induction on k). By (4.61), this holds for $k = 2$. Suppose the statement holds for k. Then we have:

$$\begin{aligned} w[D_t^{k+1} \hat{Z}(0)]_w = w\Big\{ & [D_t^k \hat{Z}(0)]_w \phi(0) + k[D_t^{k-1} \hat{Z}(0)]_w \phi_t(0) \\ & + \frac{k(k-1)}{2} [D_t^{k-2} \hat{Z}(0)]_w \phi_{tt}(0) \Big\}_w. \end{aligned} \tag{4.71}$$

Here

$$\begin{aligned} \phi(0) &= \tau = c \in w^{-(n+1)} + \delta c w^{-s} + \cdots, \\ \phi_t(0) &= -c^2 \delta i (n+1-s) w^{-(n+1+s)} - c^2 \delta^2 i (n+1-s) w^{-2s} + \cdots \end{aligned} \tag{4.72}$$

and

$$\phi_{tt}(0) = 2c^3 \epsilon \delta^2 i^2 (n+1-s)^2 w^{-(n+1+2s)} + \cdots.$$

By hypothesis the δ^2 term of the last complex component of $w[D_t^k \hat{Z}(0)]_w$ is given by (4.69).

From the $\epsilon c w^{-(n+1)}$ term of $\phi(0)$, we see that there is, by the induction hypothesis, a contribution to the δ^2 term of the last complex component of $w[D_t^k \hat{Z}(0)]_w \phi(0)$ of the form:

$$\frac{k(k-1)}{2} i^k \delta^2 \epsilon^{k-1} c^{k+1} \rho_k R_m w^{\xi}, \tag{4.73}$$

where $\xi := 1 + m - (k-1)(n+1) - 2s$.

From Lemma 4.7, we know the leading δ-linear term of $w[D_t^k \hat{Z}(0)]_w$, yielding

$$k\delta^2 i^k \epsilon^{k-1} c^{k+1} [m-n] \cdot \ldots \cdot [1+m-(k-1)(n+1)] \cdot [1+m-(k-1)(n+1)-s] R_m w^\xi. \tag{4.74}$$

We now consider the leading δ^2 term of the last complex component of $kw[D_t^{k-1} \hat{Z}(0)]_w \phi_t(0)$. Since $\phi_t(0)$ has only δ-linear or δ^2-terms, the δ^2 term of $w[D_t^{k-1} \hat{Z}(0)]_w \phi_t(0)$ does not contribute. We only have a contribution from the δ^0 and δ terms which yield:

$$\begin{aligned} &-ki^k c^{k+1} \epsilon^{k-1} \delta^2 [m-n] \cdot \ldots \cdot [1+m-(k-1)(n+1)] \cdot (n+1-s) R_m w^\xi \\ &- k(k-1) i^k c^{k+1} \epsilon^{k-1} \delta^2 [m-n] \cdot \ldots \cdot [1+m-(k-2)(n+1)] \\ &\cdot [1+m-(k-2)(n+1)-s] \cdot (n+1-s) R_m w^\xi. \end{aligned} \tag{4.75}$$

We must finally consider $\frac{k(k-1)}{2} w[D_t^{k-2} \hat{Z}(0)]_w \phi_{tt}(0)$. Here, since $\phi_{tt}(0)$ only has a δ^2-leading term, only the δ^0-term of the last complex component of $w[D_t^{k-2} \hat{Z}(0)]_w \phi_{tt}(0)$ contributes, namely:

$$k(k-1) i^k c^{k+1} \epsilon^{k-1} \delta^2 [m-n] \cdot \ldots \cdot [1+m-(k-2)(n+1)] \cdot (n+1-s) R_m w^\xi. \tag{4.76}$$

Now, wonderfully, the second term of (4.75) + (4.76) equals

$$-k(k-1) i^k c^{k+1} \epsilon^{k-1} \delta^2 [m-n] \cdot \ldots \cdot [1+m-(k-1)(n+1)] \cdot (n+1-s) R_m w^\xi. \tag{4.77}$$

Now let us consider the sum of (4.77) + (4.73). Equation (4.77) can be written as

$$\frac{-k(k-1)}{2} i^k c^{k+1} \epsilon^{k-1} \delta^2 [m-n] \cdot \ldots \cdot [1+m-(k-1)(n+1)] \cdot [2(n+1)-2s] R_m w^\xi, \tag{4.78}$$

and (4.73) is

$$\frac{k(k-1)}{2} i^k c^{k+1} \epsilon^{k-1} [m-n] \cdot \ldots \cdot [1+m-(k-1)(n+1)] \cdot [1+m-(k-2)(n+1)-2s] R_m w^\xi. \tag{4.79}$$

This sum equals

$$\frac{k(k-1)}{2} i^k c^{k+1} \epsilon^{k-1} \delta^2 [m-n] \cdot \ldots \cdot [1+m-(k-1)(n+1)] \cdot [1+m-k(n+1)] R_m w^\xi, \tag{4.80}$$

and for the remaining two terms, the sum of the first term of (4.75) + (4.74) is

$$ki^k c^{k+1} \epsilon^{k-1} \delta^2 [m-n] \cdot \ldots \cdot [1+m-(k-1)(n+1)] \cdot [1+m-k(n+1)] R_m w^\xi \tag{4.81}$$

and the sum of (4.81) and (4.80) is

$$\frac{k(k+1)}{2} i^k c^{k+1} \epsilon^{k-1} \delta^2 [m-n] \cdot \ldots \cdot [1+m-k(n+1)] R_m w^{\xi}.$$

Now multiplying by ξ (i.e. taking the derivative) yields the lemma. □

Using formula (4.66) and the independence from the first complex components, we are ready to prove Proposition 4.9. Noting that there is neither a δ nor a δ^2 pole in

$$\begin{aligned} &[D_t^{L/2-1} \hat{Z}(0)]_w \phi(0) + (L/2-1)[D_t^{L/2-2} \hat{Z}(0)]_w \phi_t(0) \\ &+ \frac{1}{2}\left(\frac{L}{2}-1\right)\left(\frac{L}{2}-2\right)[D_t^{L/2-3} \hat{Z}(0)]_w \phi_{tt}(0) \end{aligned} \tag{4.82}$$

we apply Lemma 4.8 and multiply the δ^2 zero term of $w[D_t^{L/2} \hat{Z}(0)]_w$ by the δ^0 pole. Applying Lemma 4.10 we see that the δ^0 term of $w[D_t^{L/2} \hat{Z}(0)]_w$ is

$$\begin{aligned} &\frac{L/2(L/2-1)}{2} c^{L/2} \epsilon^{L/2-2} \delta^2 i^{L/2} [m-n] \cdot \ldots \\ &\cdot [1+m-(L/2-1)(n+1)](n+1-s) R_m w^{v}, \\ &v = 1+m-(L/2-2)(n+1) - 2s = n+1-s, \end{aligned} \tag{4.83}$$

and the δ^0 pole of (4.82) is

$$i^{L/2-1} c^{L/2} \epsilon^{L/2} [m-n] \cdot \ldots \cdot [1+m-(L/2-1)(n+1)] R_m w^{1+m-L/2(n+1)}. \tag{4.84}$$

Now we have

$$1+m-L/2(n+1) = \frac{1}{2}\{r-(n+1)\} = -\frac{1}{2}(n+1-r),$$

but

$$s = (n+1+r)/2, \qquad n+1-s = (n+1-r)/2.$$

Thus

$$1+m-L/2(n+1) = -(n+1-s).$$

Hence, the first term of (4.66) yields the integral

$$\frac{2(L-1)!}{(L/2-1)!(L/2-1)!} \operatorname{Re} \int (L/2-1) c^L i^{L-1} \delta^2 \epsilon^{L-2} \mu \frac{R_m^2}{w} \, dw \tag{4.85}$$

where

$$\mu = [m-n]^2 \cdot \ldots \cdot [1+m-(L/2-1)(n+1)]^2 (n+1-s). \tag{4.86}$$

What about the second term of (4.66)?

First we consider the term

$$\frac{(L-1)!}{(L/2-1)!(L/2-1)!} \operatorname{Re} \int_{S^1} w[D_t^{L/2-1} \hat{Z}(0)]_w \cdot [D_t^{L/2-1} \hat{Z}(0)]_w \phi_t(0) \, dw. \tag{4.87}$$

Again, since no δ^0-terms of $\phi_t(0)$ contribute, we need to consider only the δ and δ^0-terms of

$$w[D_t^{L/2-1}\hat{Z}(0)]_w. \tag{4.88}$$

The δ-linear term of the last complex component of (4.88) yields a contribution to the integrand of (4.87)

$$-2\delta^2(L/2-1)c^L i^{L-1}\epsilon^{L-2}\Theta_1 R_m w^{-1} - \delta^2 c^L i^{L-1}\epsilon^{L-2}\Theta_2 R_m w^{-1} + \cdots; \tag{4.89}$$

$$\begin{aligned}\Theta_1 &:= [m-n]^2 \cdot \ldots \cdot [1+m-(L/2-2)(n+1)]^2 \cdot [1+m-(L/2-1)(n+1)]\\ &\quad \cdot [1+m-(L/2-1)(n+1)-s](n+1-s);\\ \Theta_2 &:= [m-n]^2 \cdot \ldots \cdot [1+m-(L/2-1)(n+1)]^2(n+1-s).\end{aligned}$$

Now only the δ^0-terms of

$$w[D_t^{L/2-1}\hat{Z}(0)]_w \quad \text{and} \quad w[D_t^{L/2-2}\hat{Z}(0)]_w$$

contribute to the δ^2-term of the last complex component of

$$(L/2-1)w[D_t^{L/2-1}\hat{Z}(0)]_w \cdot [D_t^{L/2-2}\hat{Z}(0)]_w \phi_{tt}(0)$$

and this is

$$\begin{aligned}&2(L/2-1)\delta^2 c^L i^{L-1}\epsilon^{L-2}\Theta_3 R_m w^{-1};\\ &\quad \Theta_3 = [m-n]^2 \cdot \ldots \cdot [1+m-(L/2-2)(n+1)]^2\\ &\qquad \cdot [1+m-(L/2-1)(n+1)](n+1-s)^2.\end{aligned} \tag{4.90}$$

However, the first term of (4.89) + (4.90) equals

$$\begin{aligned}&-2(L/2-1)\delta^2 c^L i^{L-1}\epsilon^{L-2}[m-n]^2 \cdot \ldots\\ &\quad \cdot [1+m-(L/2-1)(n+1)]^2(n+1-s).\end{aligned} \tag{4.91}$$

Again the Lth derivative does not depend on the first complex components. For $L > 6$ this follows as in Lemma 4.3. We leave $L = 6$ to the reader.

Adding (4.91), the second term of (4.89) and (4.85) proves (4.68) and thus Proposition 4.9.

The reader should also note that formula (4.68) agrees with formula (4.65), for the case $L = 4$.

Chapter 5
The Second Main Theorem: Exceptional Branch Points; The Condition $k > l$

In this and the next chapter we want to prove our *Second Main Theorem*, namely the following result:

Theorem 5.1 *Suppose that Γ is a closed rectifiable Jordan curve in $\mathbb{R}^3$, and let $\hat{X} \in \mathcal{C}(\Gamma)$ be a minimal surface having $w_0 \in B$ as an exceptional branch point. Then $\hat{X}$ is not a C^0 relative minimizer of A in $\mathcal{C}(\Gamma)$.*

Before we turn to the proof we make the following

Remark 5.1 If Γ is a planar curve then no minimal surface $\hat{X} \in \mathcal{C}(\Gamma)$ has interior branch points.

Proof We may assume that Γ lies in the x^1, x^2-plane, identified with $\mathbb{C}$, and by the maximum principle we obtain

$$\hat{X}(w) = (\hat{X}^1(w), \hat{X}^2(w), 0).$$

(See Sect 4.11 of Dierkes, Hildebrandt and Sauvigny [1].) It follows that $f := \hat{X}^1 + i\hat{X}^2$ provides a strictly conformal or anticonformal mapping $w \mapsto f(w)$ from B onto the inner domain Ω of Γ in $\mathbb{C}$; in particular we have $f'(w) \neq 0$ for $w \in B$, and therefore $\hat{X}_w(w) \neq 0$. □

Remark 5.2 Clearly a minimal surface $\hat{X} \in \mathcal{C}(\Gamma)$ is planar if and only if Γ is planar. *Therefore it suffices to prove Theorem 5.1 under the additional hypothesis that the exceptional branch point w_0 is the origin and $\hat{X} \in \mathcal{C}(\Gamma)$ is a nonplanar minimal surface in normal form at $w = 0$.*

In the following discussion we shall suppose this situation, but dispense with the assumption $\hat{X} \in \mathcal{C}(\Gamma)$; instead we assume $\hat{X} \in C^\infty(\overline{B}, \mathbb{R}^3)$. To formalize these requirements we state

A. Tromba, *A Theory of Branched Minimal Surfaces*,
Springer Monographs in Mathematics,
DOI 10.1007/978-3-642-25620-2_5,

Condition (N) *Let $\hat{X} \in C^\infty(\overline{B}, \mathbb{R}^3)$ be a nonplanar minimal surface in normal form at the exceptional branch point $w = 0$ of order n and index m, $m > n \geq 1$, i.e.*

$$m + 1 \equiv 0 \bmod (n + 1).$$

From here on, up to and including Theorem 5.10, we assume that Condition (N) is satisfied.

Consider the Taylor expansion of $\hat{X}_w$ on B at $w = 0$:

$$\hat{X}_w(w) = \left(\sum_{j=1}^{\infty} A_j w^{n+j-1}, \sum_{j=m}^{\infty} R_j w^j \right), \tag{5.1}$$

$$A_j \in \mathbb{C}^2,\ R_j \in \mathbb{C},\ A_1 \neq 0,\ R_m \neq 0,\ A_1 \cdot A_1 = 0,$$
$$A_j = \lambda_j A_1 \quad \text{for } j = 1, \ldots, 2m - 2n,$$
$$A_1 \cdot A_{2m-2n+1} = -\frac{1}{2} R_m^2.$$

Then

$$w \hat{X}_w(w) = \left(\sum_{j=1}^{\infty} A_j w^{n+j}, \sum_{j=m}^{\infty} R_j w^{j+1} \right). \tag{5.2}$$

Because of $n + j = (n + 1) + (j - 1)$ we have

$$(n + j) \not\equiv 0 \bmod (n + 1) \Leftrightarrow (j - 1) \not\equiv 0 \bmod (n + 1).$$

Suppose that

$$A_j = 0 \quad \text{for all } j \in \mathbb{N} \text{ with } (j - 1) \not\equiv 0 \bmod (n + 1). \tag{5.3}$$

This implies

$$R_j = 0 \quad \text{for all } j \geq m \text{ with } (j + 1) \not\equiv 0 \bmod (n + 1). \tag{5.4}$$

If this were not the case we consider the smallest integer $\nu > m$ such that $(\nu + 1) \not\equiv 0 \bmod (n + 1)$ and $R_\nu \neq 0$. Since $(m + 1) \equiv 0 \bmod (n + 1)$, it follows that $(m + 1) + (\nu + 1) = 2 + m + \nu \not\equiv 0 \bmod (n + 1)$. Consider the equation

$$0 = w^2 \hat{X}_w \cdot \hat{X}_w = [w^2 \hat{X}_w^1 \hat{X}_w^1 + w^2 \hat{X}_w^2 \hat{X}_w^2] + w^2 \hat{X}_w^3 \hat{X}_w^3. \tag{5.5}$$

The expression $\psi(w) := [\ldots]$ in (5.5) is a sum of terms $A_j \cdot A_s w^{n+j+n+s}$; thus, by virtue of (5.3), the power w^p with

$$p = n + j + n + s = 2n + j + s = 2(n + 1) + (j - 1) + (s - 1)$$

can only have a coefficient different from zero if $p \equiv 0 \bmod (n + 1)$. By (5.5) it follows that also $w^2 X_w^3 X_w^3$ is a sum of the form $\Sigma c_p w^p$ with $c_p = 0$ if $p \not\equiv 0 \bmod (n + 1)$. On the other hand, for $p_1 = (m + 1) + (\nu + 1)$ we have $c_{p_1} = 2 R_m R_\nu \neq 0$ and $p_1 \not\equiv 0 \bmod (n + 1)$, a contradiction.

Thus (5.3) implies (5.4), which means that, in (5.2), only those w-powers can appear whose exponents are divisible by $(n + 1)$, and we have obtained:

Observation 5.1

(i) If (5.3) holds then $w\hat{X}_w(w)$ is of the form

$$w\hat{X}_w(w) = \sum_{j=1}^{\infty} F_j w^{j(n+1)}, \tag{5.6}$$

$F_j \in \mathbb{C}^3$ with $F_1 = (A_1, 0)$, and

$$\hat{X}(w) = X_0 + \operatorname{Re} \sum_{j=1}^{\infty} \tilde{F}_j w^{j(n+1)} \tag{5.7}$$

with $X_0 \in \mathbb{R}^3$ and $\tilde{F}_j = F_j / j(n+1)$.

(ii) If (5.3) does not hold, there is an index $j_0 \in \mathbb{N}$ with $j_0 - 1 \not\equiv 0 \mod (n+1)$ such that $A_{j_0} \neq 0$ as well as $A_j = 0$ for all $j \in \mathbb{N}$ with $j < j_0$ and $j - 1 \not\equiv 0 \mod (n+1)$.

We now define two "indices" $k, l \in \overline{\mathbb{N}} := \mathbb{N} \cup \{\infty\}$ as follows:

Definition 5.1 If there is an integer j with $(j-1) \not\equiv 0 \mod (n+1)$, $A_j \neq 0$, and $A_1 \cdot A_j = 0$, then k is the least integer of this kind; otherwise we set $k = \infty$.

Secondly, if there is an integer j with $(j-1) \not\equiv 0 \mod (n+1)$, $A_j \neq 0$, $A_1 \cdot A_j \neq 0$, then l is the least integer of this kind.

Observation 5.2

(i) If $k = l = \infty$ then we have (5.3), therefore (5.4) and consequently (5.6). Otherwise $k \neq l$, and at least one of them is finite.

(ii) *We have* $l > 2m - 2n + 1$.

Proof of (ii) Since $A_1 \cdot A_j = 0$ for $1 \leq j \leq 2m - 2n$ and $(2m - 2n + 1) - 1 = 2(m+1) - 2(n+1) \equiv 0 \mod (n+1)$, it follows that $l > 2m - 2n + 1$. □

Observation 5.3 We have

$$A_k \cdot A_{2m-2n+1} \neq 0 \quad \text{if } k < \infty,$$

and

$$A_1 \cdot A_l \neq 0 \quad \text{if } l < \infty.$$

Proof From $k < \infty$ and $A_k \cdot A_1 = 0$, $A_k \neq 0$ we infer

$$A_k = \lambda_k A_1 \quad \text{with } \lambda_k \neq 0.$$

Since $A_1 \cdot A_{2m-2n+1} = -\frac{1}{2} R_m^2 \neq 0$, it follows that

$$A_k \cdot A_{2m-2n+1} = -\frac{1}{2} \lambda_k R_m^2 \neq 0.$$

Secondly, $l < \infty$ implies $A_1 \cdot A_l \neq 0$ by the definition of l. □

Suppose now that we are not in the situation when $k = l = \infty$. Then $k \neq l$, and so at least one of the "indices" k, l is finite. We distinguish three different cases:

$$\begin{aligned} &\text{case (A):} \quad 2m + 2 + (k-1) < 2(n+1) + (l-1);\\ &\text{case (B):} \quad 2m + 2 + (k-1) > 2(n+1) + (l-1);\\ &\text{case (C):} \quad 2m + 2 + (k-1) = 2(n+1) + (l-1) =: \gamma. \end{aligned}$$

Observation 5.4 If not $k = l = \infty$ we have

$$\begin{aligned} &k < \infty,\ l \leq \infty && \text{in case (A)},\\ &l < \infty,\ k \leq \infty && \text{in case (B)},\\ &k < \infty \text{ and } l < \infty && \text{in case (C)}. \end{aligned}$$

Consider now the expressions

$$\begin{aligned} \psi(w) &:= w^2 \hat{X}^1_w(w) \hat{X}^1_w(w) + w^2 \hat{X}^2_w(w) \hat{X}^2_w(w),\\ \chi(w) &= w^2 \hat{X}^3_w \hat{X}^3_w(w) \end{aligned}$$

and inspect the term with the lowest w-power in $\psi(w)$ whose exponent is not divisible by $(n+1)$. By Observations 5.3 and 5.4 it is

$$\begin{aligned} &T(w) := 2A_k \cdot A_{2m-2n+1} w^{2m+2+(k-1)} && \text{in case (A)},\\ &T(w) := 2A_1 \cdot A_l w^{2n+2+(l-1)} && \text{in case (B)},\\ &T(w) = c_\gamma \cdot w^\gamma && \text{in case (C)} \end{aligned}$$

provided that

$$c_\gamma := 2(A_k \cdot A_{2m-2n+1} + A_1 \cdot A_l) \neq 0.$$

We distinguish the two subcases (C1) and (C2) of (C), defined by

$$(\text{C1}) : c_\gamma \neq 0; \qquad (\text{C2}) : c_\gamma = 0.$$

By (5.5) we have $\psi(w) + \chi(w) \equiv 0$ on B, and so $T(w)$ in $\psi(w)$ must be compensated by a term $T^*(w)$ in $\chi(w)$ in cases (A), (B), and (C1):

$$T(w) + T^*(w) = 0.$$

The term $T^*(w)$ has to be of the form

$$T^*(w) = 2R_m R_\nu w^{2+m+\nu} \quad \text{with } R_\nu \neq 0$$

where ν is the smallest integer $j > m$ such that $R_j \neq 0$ and $(j+2) \neq 0 \bmod (n+1)$. Thus we make

Observation 5.5 If not $k = l = \infty$ then:

$$\begin{aligned} &\text{in case (A):} && 2 + m + \nu = 2(m+1) + (k-1),\ k < \infty,\ R_\nu \neq 0;\\ &\text{in case (B):} && 2 + m + \nu = 2(n+1) + (l-1),\ l < \infty,\ R_\nu \neq 0;\\ &\text{in case (C1):} && 2 + m + \nu = 2(m+1) + (k-1) = 2(n+1) + (l-1) \text{ and}\\ & && k < \infty,\ l < \infty,\ R_\nu \neq 0. \end{aligned}$$

In case (C2) the smallest integer ν introduced above may or may not exist; in the second case we set $\nu := \infty$. Then we have

$$\text{in case (C2):}\quad 2+m+\nu > 2(m+1)+(k-1) = 2(n+1)+(l-1).$$

Now we introduce the integer $\overline{m}$ as ν in cases (A), (B), and (C1), while in case (C2), ν is not suited to define $\overline{m} \in \mathbb{N}$; instead we set $\overline{m} := m+(k-1)$ in this case. In other words we have the following

Definition 5.2 If not $k = l = \infty$ then $\overline{m} \in \mathbb{N}$ is defined by

$$\begin{aligned} 2+m+\overline{m} &:= 2(m+1)+(k-1) && \text{in case (A)},\\ 2+m+\overline{m} &:= 2(n+1)+(l-1) && \text{in case (B)},\\ 2+m+\overline{m} &:= 2(m+1)+(k-1) = 2(n+1)+(l-1) && \text{in case (C)}. \end{aligned}$$

Then the preceding discussion leads to

Observation 5.6 If not $k = l = \infty$ then we have

$$\begin{aligned} &\text{in case (A):}\quad k<\infty,\ l\le\infty,\ \overline{m}=\nu<\infty,\ R_{\overline{m}}\ne 0;\\ &\text{in case (B):}\quad l<\infty,\ k\le\infty,\ \overline{m}=\nu<\infty,\ R_{\overline{m}}\ne 0;\\ &\text{in case (C):}\quad k<\infty,\ l<\infty,\ \overline{m}<\infty. \end{aligned}$$

Furthermore, in case (C)

$$R_{\overline{m}} = 0 \Leftrightarrow A_k \cdot A_{2m-2n+1} + A_l \cdot A_1 = 0$$

and in any case: $R_{\overline{m}} \ne 0 \Leftrightarrow \overline{m} = \nu$.

Proposition 5.1 *Suppose that not* $k = l = \infty$. *Then* $\overline{m} \in \mathbb{N}$ *is defined, and we obtain the following*:

(i) *In all cases* (A), (B), (C) *we have*

$$(n+\ell)-(1+\overline{m}) \ge (m-n) \ge 1. \tag{5.8}$$

(ii) *If* $k \ge \ell$, *then neither* (A) *nor* (C) *can hold, i.e. we are in case* (B).
(iii) *If* (A) *or* (C) *holds, then*

$$\ell - k \ge 1+m, \tag{5.9}$$

$$1+\overline{m} \ge (n+k)+(n+1). \tag{5.10}$$

Proof (i) In case (A) or (C) we have

$$1+\overline{m} \le 2(n+1)+(\ell-1)-(1+m) = (n+\ell)-(m-n),$$

and in case (B),

$$m+\overline{m} = 2n+\ell-1,$$

whence

$$(n+\ell)-(1+\overline{m}) = m-n.$$

(ii) is obvious.

(iii) Since the branch point is exceptional we have

$$1+m \geq 2(n+1).$$

If (A) or (C) hold, then

$$2m+2+(k-1) \leq 2(n+1)+(\ell-1)$$

whence

$$(m+1)+(k-1) \leq 2(n+1)+(\ell-1)-(1+m) \leq (\ell-1),$$

and so

$$\ell-k \geq m+1.$$

Furthermore,

$$1+\overline{m}=(1+m)+(k-1) \geq 2(n+1)+(k-1)=(n+k)+(n+1). \qquad \square$$

Recall that, by assumption, $(m+1) \equiv 0 \bmod (n+1)$, and that $m > n \geq 1$. Thus we have

$$m+1=p(n+1) \quad \text{for some } p \in \mathbb{N} \text{ with } p \geq 2, \tag{5.11}$$

and in particular

$$m+1 \geq 2(n+1) \Leftrightarrow m \geq 2n+1 \geq 3. \tag{5.12}$$

Furthermore we have $\overline{m} > m$ and $\overline{m}+1 \not\equiv 0 \bmod (n+1)$. Therefore,

$$(1+\overline{m})-(1+m)>0 \quad \text{and} \quad (1+\overline{m})-(1+m) \not\equiv 0 \bmod (n+1).$$

Thus there are integers Γ and s with $\Gamma \geq 0$ and $0 < s < n+1$ such that

$$(1+\overline{m})-(1+m)=\Gamma(n+1)+s,$$

and we obtain

Proposition 5.2 *Suppose that not $k=l=\infty$. Then there are integers Γ and s with $\Gamma \geq 0$ and $0 < s < n+1$ such that*

$$(\overline{m}+1)=(m+1)+\Gamma(n+1)+s. \tag{5.13}$$

Proposition 5.3 *The number*

$$L:=\frac{2(m+1)}{(n+1)} \tag{5.14}$$

is an even integer with $L \geq 4$.

Proof Because of (5.11) we have $L=2p \geq 4$. $\square$

Definition 5.3 If not $k = l = \infty$ we distinguish the three cases

(I) $s = \frac{1}{2}(n+1)$,
(II) $0 < s < \frac{1}{2}(n+1)$,
(III) $\frac{1}{2}(n+1) < s < (n+1)$.

In cases (I) and (II), i.e. for $0 < s \leq \frac{1}{2}(n+1)$, we set

$$\overline{L} := L + 2\Gamma + 1; \tag{5.15}$$

here $\overline{L}$ is an odd integer ≥ 5.

In case (III) we define $\overline{L}$ as

$$\overline{L} := L + 2\Gamma + 2. \tag{5.16}$$

From (5.13) and (5.14) it follows

$$2(\overline{m} + 1) = L(n+1) + 2\Gamma(n+1) + 2s$$

whence

$$\frac{2(\overline{m}+1)}{(n+1)} = (L + 2\Gamma) + \frac{2s}{n+1}, \quad 0 < s < n+1.$$

Here $L + 2\Gamma$ is even, and $0 < \frac{2s}{n+1} < 2$. Hence $\frac{2s}{n+1}$ is an integer if and only if $s = \frac{n+1}{2}$. Thus we obtain

Proposition 5.4 *Exactly in case* (I), *the quotient* $2(\overline{m}+1)/(n+1)$ *is an integer; in fact, it is the odd integer* $\overline{L}$. *In other words*:

$$\overline{L} = \frac{2(\overline{m}+1)}{(n+1)} \quad \text{if and only if } s = \frac{1}{2}(n+1). \tag{5.17}$$

Proposition 5.5 *If not* $k = l = \infty$ *we have*

$$k - 1 = \Gamma(n+1) + s \tag{5.18}$$

in case (A) *or* (C), *and*

$$\ell - 1 = 2(m+1) + (\Gamma - 2)(n+1) + s \tag{5.19}$$

in case (B).

Proof In case (A) or (C) we have

$$2 + m + \overline{m} = 2(m+1) + (k-1)$$

whence

$$(\overline{m}+1) = (m+1) + (k-1) \overset{(5.13)}{=} (m+1) + \Gamma(n+1) + s,$$

which yields (5.13).

If (B) holds then

$$\begin{aligned} 2(n+1)+(\ell-1) &= (m+1)+(\overline{m}+1) \\ &\overset{(5.13)}{=} 2(m+1)+\Gamma(n+1)+s. \end{aligned}$$ □

From Definition 5.3 we immediately infer

Proposition 5.6 *Suppose that not $k=l=\infty$. Then we have:*

(i) *If $s \leq \frac{1}{2}(n+1)$ then $\overline{L}$ is odd, and*

$$\frac{\overline{L}-1}{2} = \frac{L}{2} + \Gamma. \tag{5.20}$$

(ii) *If $s > \frac{1}{2}(n+1)$ then $\overline{L}$ is even, and*

$$\frac{\overline{L}}{2} - 1 = \frac{L}{2} + \Gamma. \tag{5.21}$$

Suppose that not $k=l=\infty$.

In Chap. 4 we have seen that a minimal surface $\hat{X} \in \mathcal{C}(\Gamma)$ cannot be a weak relative minimizer of D in $\mathcal{C}(\Gamma)$ if $w=0$ is a non-exceptional branch point of X. The analogous statement is not true if $w=0$ is exceptional. In a first example we will indicate why our method can fail, and then a second example will show that the method has to fail.

Primary Example To give the reader some insight to the methods we will be using, let us consider as an example the special case $k=2, \ell=6, n=1, m=3$. Here we are in case (C) with

$$2m+2+(k-1)=2(n+1)+(\ell-1), \qquad \overline{m}=m+(k-1)=4.$$

Consider the expansion of $w\hat{X}_w(w)$, which is given by

$$w\hat{X}_w(w) = (A_1 w^2 + A_2 w^3 + \cdots + A_6 w^7 + \cdots, R_3 w^4 + R_4 w^5 + \cdots),$$

and choose the generator τ as

$$\tau := c w^{-2} + \delta c w^{-1} + \overline{c} w^2 + \overline{\delta}\overline{c} w^1$$

where $\delta \in \mathbb{C}$ denotes an arbitrary parameter. Note that

$$A_2 = \lambda A_1$$

since $k=2$. Furthermore we choose $\phi_t(0)$ as

$$\begin{aligned} \phi_t(0) &:= -i\lambda c^2 w^{-3} - i c^2 \delta w^{-3} + i\overline{\lambda}\overline{c}^2 w^3 + i\overline{\delta}\overline{c}^2 w^3 \\ &= -i(\lambda+\delta) c^2 w^{-3} + \cdots. \end{aligned}$$

Since $1+m = 2(1+n)$ (and so $w=0$ is an exceptional branch point), one checks that $E^{(j)}(0)=0$ for $1 \leq j \leq 4$. Let us now consider $E^{(5)}(0)$. Here $\Gamma=0, s=$

$1, L = 4$ and $\overline{L} = L + 2\Gamma + 1 = 5$. A straightforward computation yields

$$E^{(5)}(0) = 12 \operatorname{Re} \int_{S^1} c^5 \kappa \frac{dw}{w}$$

with

$$\kappa := (3R_4 + 4\delta R_3 - \lambda R_3)^2 - 4(3R_4 + 4\delta R_3 - \lambda R_3) 2R_3(\lambda + \delta) \\ +16R_3^2(\lambda + \delta)^2, \quad R_3 \neq 0.$$

We note that the last term in this expression stems from the first complex components of $\hat{Z}_w(0)$, and so $E^{(5)}(0)$ may depend on these components. However, if we set $\delta := -\lambda$, we obtain $\kappa = (3R_4 - 5R_3\lambda)^2$ and therefore $\kappa \neq 0$ if $\lambda \neq 3R_4/(5R_3)$. In the second case $E^{(5)}(0)$ can be made negative while in the first case $E^{(5)}(0) = 0$, and so no general statement can be made. In fact, this observation indicates a principal drawback of our method as we shall see from the next example that has been communicated to us by F. Tomi. To explain this example we need two observations, pointed out by Tomi.

Observation 5.7 Let G be an open, convex set in $\mathbb{R}^3$, and ω be a closed smooth 2-form on G. Suppose $F_1, F_2 \in C^1(\overline{B}, \mathbb{R}^3)$ satisfy $F_1|\partial B = F_2|\partial B$. Then

$$\int_B F_1^* \omega = \int_B F_2^* \omega.$$

Proof There is some smooth 1-form η on G such that $\omega = d\eta$. Then

$$\int_B F_j^* \omega = \int_B F_j^* d\eta = \int_B d(F_j^* \eta) = \int_{\partial B} F_j^* \eta, \quad j = 1, 2,$$

and

$$\int_{\partial B} F_1^* \eta = \int_{\partial B} F_2^* \eta. \qquad \square$$

Consider a minimal graph M over a convex domain Ω in $\mathbb{R}^2$,

$$M = \{x = (x^1, x^2, x^3) \in \mathbb{R}^3 : x^3 = \psi(x^1, x^2), (x^1, x^2) \in \Omega\}$$

and let $n = (n_1, n_2, n_3) : G \to S^2 \subset \mathbb{R}^3$ be a unit vector field on the convex cylinder $G := \Omega \times \mathbb{R} \subset \mathbb{R}^3$ which is obtained by vertical translation from the normal field of M. Then we have $\operatorname{div} n = 0$ on G whence the 2-form

$$\omega := n_1 dx^2 \wedge dx^3 + n_2 dx^3 \wedge dx^1 + n_3 dx^1 \wedge dx^2$$

is closed. This leads to

Observation 5.8 Let $\hat{X} \in C^\infty(\overline{B}, \mathbb{R}^3)$ be a minimal surface with no branch point on ∂B such that $\hat{X}(\overline{B}) \subset M$ where M is a minimal graph as considered before. Furthermore, let $\hat{Y} \in C^1(\overline{B}, \mathbb{R}^3)$ be an arbitrary surface with $Y(\overline{B}) \subset G$ such that the boundary values $X := \hat{X}|_{\partial B}, Y := \hat{Y}|_{\partial B}$ satisfy (i) $X = Y$, or more generally

(ii) $Y = X \circ \varphi$ where φ is a diffeomorphism of $S^1 = \partial B$ that is homotopic to the identity id_{S^1}. Then,

$$D(\hat{X}) = A(\hat{X}) \le A(\hat{Y}) \le D(\hat{Y}).$$

Proof Because of $\hat{X}(\overline{B}) \subset M$ we can assume that the Gauss map $N : \overline{B} \to S^2 \subset \mathbb{R}^3$ of $\hat{X}$ satisfies $N = n \circ \hat{X}$, whence

$$A(\hat{X}) = \int_B \hat{X}^* \omega.$$

Furthermore we infer from $|n| = 1$ that

$$\int_B \hat{Y}^* \omega \le A(\hat{Y}),$$

and in case (i) it follows

$$\int_B \hat{X}^* \omega = \int_B \hat{Y}^* \omega$$

on account of Observation 5.7. This implies $A(\hat{X}) \le A(\hat{Y})$. In case (ii) we can extend $\varphi : S^1 \to S^1$ to a diffeomorphism ϕ of B onto itself, and so the preceding reasoning yields

$$A(\hat{X}) \le A(\hat{Y} \circ \phi).$$

However, $A(\hat{Y} \circ \phi) = A(\hat{Y})$, and so we obtain the desired inequality. □

Now we turn to the

Secondary Example Let $F : \mathbb{C} \to \mathbb{R}^3$ be a rescaling of Enneper's surface, given by

$$\hat{F}(z) = \frac{1}{2} \operatorname{Re}\left(z - \frac{1}{3 \cdot 64} z^3, i \left(z + \frac{1}{3 \cdot 64} z^3 \right), \frac{1}{8} z^2 \right), \quad z \in \mathbb{C}.$$

We introduce the branch points $w = 0$ and $w = -\frac{3}{2}$ in $\hat{F} \circ f$ by means of

$$f(w) = w^2 \left(w + \frac{3}{2} \right)^2 = w^4 + 3w^3 + \left(\frac{9}{4} \right) w^2.$$

Setting $\hat{X} := \hat{F} \circ f|_{\overline{B}}$ we obtain a minimal surface $\hat{X} \in C^0(\overline{B}, \mathbb{R}^3)$ which has a single branch point at $w = 0$ of order $n = 1$ and index $m = 3$; yet because of the branch point $w = -3/4$ outside of $\overline{B}$, the expansion of $\hat{X}(w)$ does not proceed with powers of w^2 alone, but odd powers of w appear. *Therefore the branch point $w = 0$ of $\hat{X}$ is not globally analytically false.*

On the other hand, by a corresponding property of Enneper's surface (cf. Nitsche [1], §92), it follows that $\hat{X}(\overline{B})$ lies on a minimal graph M as described before, and so *we have the minimum property stated in Observation 5.8.*

Furthermore one computes that

$$\hat{X}_w(w) = \frac{1}{2}\left(\frac{9}{2}w + 9w^2 + 4w^3 - \frac{1}{32}\left[w\left(w+\frac{3}{2}\right)\right]^5\left(2w+\frac{3}{2}\right),\right.$$
$$i\left\{\frac{9}{2}w + 9w^2 + 4w^3 + \frac{1}{32}\left[w\left(w+\frac{3}{2}\right)\right]^5\left(2w+\frac{3}{2}\right)\right\},$$
$$\left.\frac{81}{32}w^3 + \frac{5\cdot 27}{16}w^4 + \cdots\right).$$

Considering $w\hat{X}_w(w)$ it follows $A_2 = 2A_1$ and so $\lambda = 2$, $k = 2$, $m = 3$, $\overline{m} = 4$, $R_m = 81/32$, $R_{\overline{m}} = \frac{5\cdot 27}{16}$ i.e.

$$3R_4 - 5R_3\lambda = 0.$$

Thus we are in the situation of our Primary Example where *the method fails to prove* $E^{(5)}(0) < 0$; we only have $E^{(5)}(0) = 0$ together with $E^{(j)}(0) = 0$, $j = 1, \ldots, 4$, and a further calculation shows $E^{(6)}(0) > 0$, in agreement with Observation 5.2.

What can we infer from Tomi's example? The following *Conjecture* might seem plausible, as it would be analogous to the First Main Theorem: *If $\hat{X}$ is a nonplanar minimal surface with $w = 0$ as an exceptional branch point of order n and $D(\hat{X}) \le D(\hat{Y})$ for all smooth $\hat{Y} = \overline{B} \to \mathbb{R}^3$ with*

$$\|\hat{Y} - \hat{X}\|_{C^0(\overline{B},\mathbb{R}^3)} \ll 1$$

and such that the boundary values are related by $Y = X \circ \varphi$, φ a C^∞-diffeomorphism of ∂B onto itself, then $\hat{X}$ is globally analytically false, i.e. there is a minimal surface $\hat{X}_0 \in C^\infty(\overline{B}, \mathbb{R}^3)$ such that $\hat{X}(x) = \hat{X}_0(w^{n+1})$ on $\overline{B}$.

Tomi's example shows that there is an (absolute) minimizer $\hat{X}$ of A (which then also is a D-minimizer) compared with all $\hat{Y} \in C^\infty(\overline{B}, \mathbb{R}^3)$ such that $\hat{Y} = \hat{X} \circ \varphi$, $\varphi : \partial B \to \partial B$ a diffeomorphism with $\varphi \sim \mathrm{id}_{\partial B}$, such that the power series expansion about the branch point $w = 0$ of order n contains powers w^p with $p \not\equiv 0$ mod $(n+1)$. *Hence the Conjecture is not true*, and so we have to give up the hope that for exceptional branch points one can prove a result that is as strong as that for non-exceptional branch points derived in Chap. 4.

Observation 5.9 Later on we will be reducing energy of a minimal surface in a neighbourhood of a branch point. So what can we infer from Tomi's example locally? Consider an immersed minimal surface $\hat{X}$ of the form

$$\hat{X}(w) = \mathrm{Re}\left(A_1 w + \frac{1}{2\tilde{m}+1}A_{2\tilde{m}+1}w^{2\tilde{m}+1} + \cdots, \frac{1}{\tilde{m}+1}R_{\tilde{m}}w^{\tilde{m}+1} + \cdots\right)$$

with $\tilde{m} \ge 1$. In a sufficiently small neighbourhood of $w = 0$, $\hat{X}$ is absolutely area minimizing above the x^1, x^2-plane. Now we introduce a branch point $z = 0$ by composing $\hat{X}$ with φ, given by

$$\varphi(z) := z^{n+1} + \rho z^\alpha, \quad \alpha \in \mathbb{N},\ \alpha = n + k,\ \alpha \not\equiv 0 \bmod (n+1).$$

A brief computation shows that $\hat{X} \circ \varphi$ has the exceptional branch point $z = 0$ of order n and index m with

$$1 + m = (1 + \tilde{m})(n + 1), \qquad 1 + \overline{m} = \tilde{m}(n + 1) + \alpha.$$

Thus

$$2 + m + \overline{m} = (2\overline{m} + 1)(n + 1) + \alpha.$$

Moreover, one computes that

$$l - 1 = (2\tilde{m} - 1)(n + 1) + \alpha, \quad \text{whence } l - 1 \not\equiv 0 \bmod (n + 1).$$

Then it follows

$$2(n + 1) + (l - 1) = (2\tilde{m} + 1)(n + 1) + \alpha = (2m + 2) + (k - 1).$$

Thus, for the exceptional branch point $z = 0$ of $\hat{X}(\varphi(z))$ we are in case (C). Note also that we have $k < l$ since

$$k = \alpha - n, \qquad l = (2\tilde{m} - 1)(n + 1) + \alpha + 1 > k.$$

Reversing the argument, we see that an exceptional branch point with $\infty > k > l$ cannot be an analytically false branch point. Thus, this example, together with Theorem 5.2 below indicates the importance of the *assumption*

$$k > l \quad \text{with } l < \infty$$

in order to reduce energy if $w = 0$ is an exceptional interior branch point.

In fact, ir appears impossible to calculate derivatives in the exceptional case without the assumption $k > l$. Even in this case there appears to be no simple formula for the first non-vanishing derivative.

To set the stage for the general theory, we prove a theorem for exceptionally branched minimal surfaces that is analogous to what we have proved in the non-exceptional case, but more restrictive.

Theorem 5.2 *Assume that we are in case* (I). *Then $\overline{L}$ is an odd integer given by $\overline{L} = 2(\overline{m} + 2)/(n + 1)$. Suppose also that $k > l$, $l < \infty$. Then, for an appropriate choice of $\phi(t)$, we have*

$$E^{(j)}(0) = 0 \quad \text{for } 1 \le j < \overline{L}, \qquad E^{(\overline{L})}(0) < 0.$$

Proof Since $k > \ell$, we obtain

$$n + k > n + \ell > \frac{1}{2}\overline{L}(n + 1) = 1 + \overline{m}.$$

We choose

$$\tau := cw^{-n-1} + \overline{c}w^{n+1} (= \phi(0)).$$

In forming $E^{(\alpha)}(0)$ for $\alpha \le \frac{1}{2}(\overline{L} - 1)$, no poles arise on the A_k-term; we may take

$$D_t^{\beta}\phi(0) = 0 \quad \text{for all } \beta \ge 1$$

and we can achieve

$$\hat{Z}_{tw}(0) = (iw\hat{X}_w\tau)_w,$$
$$[D_t^\alpha \hat{Z}(0)]_w = \{iD_t^{\alpha-1}\hat{Z}(0)_w\tau\}_w \quad \text{for } \alpha \le \frac{1}{2}(\overline{L}-1).$$

If we use the formula for $E^{(\overline{L})}(0)$, $\overline{L}$ odd (see Chap. 4), we are left with only one term, namely

$$E^{(\overline{L})}(0) = \frac{2(\overline{L}-1)}{M!M!}\operatorname{Re}\int_{S^1} w[D_t^M\hat{Z}(0)]_w \cdot [D_t^M\hat{Z}(0)]_w\tau\, dw$$
$$\text{with } M := \frac{1}{2}(\overline{L}-1).$$

Since the first two complex components of $w[D_t^M\hat{Z}(0)]_w$ are of the form

$$\text{const}\, c^M(A_j w^{\kappa_j(n+1)} + \cdots), \quad \kappa_j > 0,$$

the contribution to the above integral comes only from the product of the third component of $[D_t^M\hat{Z}(0)]_w$ with itself. Thus we obtain

$$E^{(\overline{L})}(0) = \frac{2(\overline{L}-1)}{M!M!}\operatorname{Re}\int_{S^1} i^{\overline{L}-1}c^{\overline{L}}\eta^2 R_{\overline{m}}^2 \frac{dw}{w} \tag{5.22}$$

with

$$\eta := (\overline{m}+1-(n+1))(\overline{m}+1-2(n+1))\dots(\overline{m}+1-M(n+1)). \tag{5.23}$$

Hence $E^{(\overline{L})}(0)$ can be made negative by an appropriate choice of c, provided that $k > \ell$. □

Observation 5.10 As in the non-exceptional case, of n is odd and $\overline{m}$ is even and $k > \ell$, then above we have a formula for the $(1+\overline{m})^{\text{th}}$ derivative of Dirichlet's energy, which can be made negative while all lower order derivatives vanish.

In general however, this appears not to be possible in the exceptional case, and it remains an open question whether such a formula can be achieved. Thus, in general, we need to argue indirectly, that some first non-vanishing derivative can be made negative.

Now we begin with the general discussion.

Theorem 5.3 (Normal form) *Suppose that $\overline{L}$ is odd, $k > l$, and $s < (n+1)/2$, and set*

$$\tau := \epsilon c w^{-(n+1)} + \epsilon\overline{c}w^{n+1} + \gamma, \quad \gamma \in \mathbb{R}, \tag{5.24}$$
$$D_t^\Gamma\phi(0) := \epsilon^\Gamma\left\{\delta c^{\Gamma+1}i^\Gamma w^{-s} + \overline{\delta}(\overline{c})^{\Gamma+1}(-i)^\Gamma\omega^s\right\}, \tag{5.25}$$

and choose $D_t^{\Gamma+1}\phi(0)$ as necessitated by the pole-removal technique, namely

$$D_t^{\Gamma+1}\phi(0) = -\epsilon^{\Gamma+1}c^{\Gamma+2}i^{\Gamma+1}(n+1-s)\delta w^{-(n+1+s)} + \cdots. \tag{5.26}$$

Assertion:

(i) *We have*

$$E^{(\overline{L})}(0) = \frac{2(\overline{L}-1)!}{(M!)^2} J_{\overline{L}} + O(\epsilon^{\overline{L}}) + O(\delta^2 \epsilon^{\overline{L}-1}) + O(\epsilon^{\overline{L}}) \tag{5.27}$$

where $J_{\overline{L}}$ *is given by*

$$J_{\overline{L}} := \mathrm{Re} \int_{S'} w[D_t^M \hat{Z}(0)]_w \cdot \Big\{ [D_t^M \hat{Z}(0)]_w \tau + \frac{2M!}{N!(\Gamma+1)!} [D_t^N \hat{Z}(0)]_w D_t^{\Gamma+1} \phi(0) \Big\}\, dw \tag{5.28}$$

with $M := (\overline{L}-1)/2$, $N := L/2 - 1$.

(ii) *Furthermore,* $E^{(j)}(0) = 0$ *for* $j = 1, \ldots, \overline{L}-1$.

(iii) *In addition, since the only non-constant generators are* $1/z^n$ *and* $1/z^s$, *it follows as in our Fundamental Computational Principles of Chap. 4,* $D_t^j \phi(0)$ *are chosen to be zero for* $1 \le j < \Gamma$, *and* $D_t^j \phi(0)$ *for* $j > \Gamma + 1$ *play no role in the computation of* $E^{(\overline{L})}(0)$. *One shows first that* $D_t^{\Gamma+2} \phi(0)$ *has poll terms too low to contribute to the* δ*-linear terms of the* $\overline{L}^{\text{th}}$ *derivative. Inductively, it then follows that all higher order derivative also do not contribute*

(iv) *Under the assumption* $k > l$, *the* ϵ^{L-1}*-term of* (5.27) *depends only on the last complex component of* $\hat{Z}_w(0)$.

The proof of this theorem is deferred until later.

The following lemma will be crucial for our reasoning.

Lemma 5.1 *The last complex component of* $w[D_t^M \hat{Z}(0)]_w$ *has a* δ*-linear term of the form*

$$\epsilon^{L/2+\Gamma-1} c^M i^M \delta \kappa R_m w^{n+1-s}. \tag{5.29}$$

If

$$\kappa - \frac{M!}{N!(\Gamma+1)!}(n+1-s)\mu \neq 0 \tag{5.30}$$

with

$$\mu = (m+1-(n+1))(m+1-2(n+1))(m+1-3(n+1))\ldots(n+1)$$

then $E^{\overline{L}}(0) \neq 0$.

Proof Apply Theorem 5.3. Then,

$$\begin{aligned} &\text{the last complex component of } w[D_t^M \hat{Z}(0)]_w \\ &= i^M c^M \{\epsilon^M k_1 R_{\overline{m}} w^s + \epsilon^{L/2+\Gamma-1} \kappa \delta R_m w^{n+1-s}\} + \cdots, \end{aligned} \tag{5.31}$$

where

$$k_1 := (\overline{m}+1-(n+1))(\overline{m}+1-2(n+1))(\overline{m}+1-3(n+1))\dots s,$$

and using (5.26), we find

$$\begin{aligned} &\text{the last complex component of } w[D_t^M \hat{Z}(0)]_w \tau \\ &\quad + \frac{2M!}{N!(\Gamma+1)!} w[D_t^N \hat{Z}(0)]_w D_t^{\Gamma+1}\phi(0) \\ &= i^M c^{M+1} \Big\{ \epsilon^{M+1} k_1 R_{\overline{m}} w^{-(n+1-s)} + \epsilon^{L/2+\Gamma} \kappa\delta R_m w^{-s} \\ &\quad - \frac{2M!\mu}{N!(\Gamma+1)!} [\epsilon^{L/2}(n+1-s)\delta R_m w^{-s}] \Big\} + \cdots. \end{aligned} \tag{5.32}$$

Therefore, the integrand of (5.28) equals

$$\begin{aligned} &2c^{\overline{L}}\epsilon^{\overline{L}-1} i^{\overline{L}-1} \Big\{ (k_1 R_{\overline{m}}) \\ &\cdot \left[\kappa\delta - \frac{M!}{N!(\Gamma+1)!}(n+1-s)\mu\delta \right] R_m \Big\} w^{-1} + O(\epsilon^L) + O(\epsilon^{L-1}\delta^2) + \cdots. \end{aligned} \tag{5.33}$$

This implies $E^{(\overline{L})}(0) \neq 0$. □

If (5.30) does not hold we cannot conclude that $E^{(\overline{L})}(0) \neq 0$. It may be that by inductive methods, one can show that in fact (5.30) is equal to zero. The inductive hypothesis is that the δ-linear term of $w[D_t^{r+k}\hat{Z}(0)]_w$, $k \geq 1$ is given by $\xi R_m w^{1+m-(k-1)(n+1)-s}$ where

$$\begin{aligned} \xi = \epsilon^{\Gamma+k-1} c^k i^k &\frac{(\Gamma+k)!}{(k-1)!(\Gamma+1)!}[m-n]\cdot\ldots \\ &\cdot [1+m-(k-1)(n+1)][1+m-(k-1)(n+1)-s]. \end{aligned}$$

Setting $k = L/2$ proves that $k = \frac{M!}{N!(\Gamma+1)!}(n+1-s)\mu$. However, we will be able to prove that if $E^{\overline{L}}(0)$ vanishes identically then $E^{(\overline{L}+1)}(0)$ can be made negative. We formulate this result as follows:

Theorem 5.4 *Let $k > l$ be satisfied. Suppose that $\overline{L}$ is odd, and assume that $E^{(\overline{L})}(0) = 0$ for all choices of τ, $D_t^\Gamma\phi(0)$, $D_t^{\Gamma+1}\phi(0)$ which ensure that $E^{(j)}(0)$ for $j = 1, \dots, \overline{L}-1$. Then $E^{(\overline{L}+1)}(0)$ can be made negative, while all lower order derivatives of $E(t)$ at $t = 0$ vanish.*

Proof Choose τ as in (5.24) and set

$$\begin{aligned} D_t^\Gamma\phi(0) &:= \epsilon^\Gamma \left\{ \delta c^{\Gamma+1} i^\Gamma w^{-s} + \overline{\delta}(\overline{c})^{\Gamma+1}(-i)^\Gamma w^s \right\}, \\ D_t^{\Gamma+1}\Phi(0) &:= \epsilon^\Gamma \left\{ \omega c^{\Gamma+2} i^{\Gamma+1} w^{-s} + \overline{\omega}(\overline{c})^{\Gamma+2}(-i)^{\Gamma+1} w^s \right\} \\ &\qquad + \text{all those terms needed to remove poles.} \end{aligned} \tag{5.34}$$

□

If we calculate $E^{(\overline{L}+1)}(0)$ for $\omega = 0$, as above, we would obtain the result that

$$E^{(\overline{L}+1)}(0) = e^{\overline{L}-1}\delta \int kc^{L+1}\frac{dw}{w} + \cdots .$$

If $k \neq 0$, we can choose c so that $E^{(\overline{L}+1)}(0) < 0$. Thus, we may assume that the δ-linear term does **not** contribute to the $(\overline{L}+1)^{\text{st}}$ derivative. We call this assumption (V).

To proceed further we need another normal form theorem whose proof we defer as well.

Theorem 5.5 (Normal form) *Suppose $k > l$. If $\overline{L}$ is odd ($\Leftrightarrow 2s < n+1$), and under assumption (V) with τ, $D_t^{\Gamma}\Phi(0)$, $D_t^{\Gamma+1}\phi(0)$ chosen as in Theorem 5.4 and $\omega = -\delta\gamma(n+1-s)$, we have:*

$$E^{(\overline{L}+1)}(0) = \frac{4\overline{L}!}{M!(M+1)!}J_{\overline{L}+1} + O(\epsilon^{\overline{L}}) + O(\epsilon^{\overline{L}-1}\delta^2) + O(\epsilon^{\overline{L}-1}\omega^2) \quad (5.35)$$

with

$$J_{\overline{L}+1} := \text{Re} \int_{S^1} w[D_t^{M+1}\hat{Z}(0)]_w \cdot \left\{[D_t^M\hat{Z}(0)]_w\tau + \frac{M!}{N!(\Gamma+1)!}[D_t^N\hat{Z}(0)]_w D_t^{\Gamma+1}\phi(0)\right\} dw,$$

$$M := \frac{1}{2}(\overline{L}-1), \quad N := L/2 - 1, \quad \textit{provided that } 2s \neq (n+1).$$

Note that the value of $E^{(\overline{L}+1)}(0)$ depends only on the last complex component of $\hat{X}_w$.

The choice of ω is made to ensure that the term in $D_t^{\Gamma+1}\phi(0)$ of order ϵ^{Γ} is used to kill a pole. By the Fundamental Computational Principles of Chap. 4, all higher order derivatives $D_t^{\beta}\phi(0)$, $\beta > \Gamma+1$ do not affect the final result.

We defer the **proof of Theorem 5.5** until later and proceed with the proof of Theorem 5.4.

Recall that

$$[D_t^M\hat{Z}(0)]_w = \left\{\text{Re}\, H\left[w(D_t^M\hat{Z}(0))_w\tau + \frac{M!}{N!(\Gamma+1)!}(D_t^N\hat{Z}(0))_w D_t^{\Gamma+1}\phi(0)\right]\right\}_w . \quad (5.36)$$

Consider the last complex component of

$$w[D_t^M\hat{Z}(0)]_w\tau + \frac{M!}{N!(\Gamma+1)!}w[D_t^N\hat{Z}(0)]_w D_t^{\Gamma+1}\phi(0); \quad (5.37)$$

the ϵ-term of order $O(\epsilon^{M+1})$ of this expression is given by

$$i^M c^{M+1}\epsilon^{M+1}k_1 R_{\overline{m}}w^{-(n+1-s)} + \cdots$$

and since, by assumption (5.30) vanishes and $1+\overline{m}=\Gamma(n+1)+(1+m)$, there are no other poles in (5.37) with a lower order ϵ. Choose

$$\omega := -\delta\gamma(n+1-s). \tag{5.38}$$

Then, as in the non-exceptional case, $E^{(\overline{L}+1)}(0)$ does not depend on $D_t^\beta \phi(0)$ for $\beta > \Gamma+1$. Moreover, with the above choice of ω, the ϵ-term of order $O(\epsilon^{L/2+\Gamma-1})$ is given by

$$c^{M+1} i^M \epsilon^{L/2+\Gamma-1} \mathcal{Z} w^{n+1-s} + \cdots, \quad \mathcal{Z} \neq 0, \tag{5.39}$$

i.e. it has a zero of order $n+1-s$ in w. Then $E^{(\overline{L}+1)}(0)$ reduces to

$$\begin{aligned} E^{(\overline{L}+1)}(0) &= \frac{4\overline{L}!}{M!(M+1)!} J_{\overline{L}+1} + O(\epsilon^{\overline{L}}) \\ &= \frac{4\overline{L}!}{M!(M+1)!} \operatorname{Re} \int_{S^1} \epsilon^{\overline{L}-1} c^{\overline{L}+1} i^{\overline{L}} (n+1-s) k_1 R_{\overline{m}} \mathcal{Z} \frac{dw}{w} \\ &\quad + O(\epsilon^{\overline{L}}) + O(\epsilon^{\overline{L}-1}\delta^2) + O(\epsilon^{\overline{L}-1}\omega^2), \end{aligned} \tag{5.40}$$

which can be made negative for sufficiently small $\epsilon > 0$ and an appropriate choice of c.

Now we turn to the case "$\overline{L}$ is even". There we are in case (III), i.e.

$$n+1 < 2s < 2n+2.$$

Here $\overline{L}$ is defined by $\overline{L} = L + 2\Gamma + 2$, and so

$$\overline{L} - 1 = L + 2\Gamma + 1 > L.$$

Consequently, if we want to achieve

$$E^{(\overline{L}-1)}(0) \neq 0, \qquad E^{(j)}(0) = 0 \quad \text{for } j = 1, \ldots, \overline{L}-2,$$

we will have $E^{(L)}(0) = 0$.

In the following discussion, the derivatives $E^{(\overline{L}-1)}(0)$ and $E^{(\overline{L})}(0)$ will play the roles of $E^{(\overline{L})}(0)$ and $E^{(\overline{L}+1)}(0)$ respectively in the preceding case where $\overline{L}$ was assumed to be odd ($\Leftrightarrow$ case (I) or (II)).

We further note that in the term corresponding to (5.32), the dominant pole term in w is now the one of order s (instead of order $(n+1-s)$). Our procedures for "$\overline{L}$ even" will parallel those for "$\overline{L}$ odd", except $\overline{L}$ is replaced by $\overline{L}-1$, and $\overline{L}+1$ by $\overline{L}$.

Theorem 5.6 (Normal form) *Suppose that $\overline{L}$ is even, $k > l$, and define τ and $D_t^\Gamma \phi(0)$ by*

$$\begin{aligned} \tau &:= \epsilon c w^{-(n+1)} + \epsilon \overline{c} w^{n+1} + \gamma, \quad \gamma \in \mathbb{R}, \\ D_t^\Gamma \phi(0) &:= \epsilon^\Gamma \{\delta c^{\Gamma+1} i^\Gamma w^{-s} + \overline{\delta} (\overline{c})^{\Gamma+1} (-i)^\Gamma w^s\}, \end{aligned}$$

and choose $D_t^{\beta}\phi(0)$ *for* $\beta \geq \Gamma + 1$ *as in* (5.26). *Then*

$$E^{(\overline{L}-1)}(0) = \frac{2(\overline{L}-1)!}{(M!)^2} J_{\overline{L}} + O(\epsilon^{\overline{L}-1}) + O(\delta^2 \epsilon^{\overline{L}-2})$$

where $J_{\overline{L}}$ *is given by* (5.28), *but* $M := \overline{L}/2 - 1$, $N := L/2 - 1$.

Now we define a new $\kappa \in \mathbb{C}$, similar to the κ defined earlier, by the observation that the last complex component of $w[D_t^M \hat{Z}(0)]_w$ (with $M = \overline{L}/2 - 1$) has a δ-linear term of the form

$$\epsilon^M c^M i^M \kappa \delta R_m w^{n+1-s} + \cdots.$$

We now obtain the analogue of the Lemma 5.1, which is proved in the same way.

Lemma 5.2 *Suppose that* $k > l$ *and*

$$\kappa - \frac{M!}{N!(\Gamma+1)!}(n+1-s)\mu \neq 0 \quad \text{with } M := (\overline{L}/2) - 1. \tag{5.41}$$

Then $E^{(\overline{L}-1)}(0) \neq 0$.

Thus we turn to the $\overline{L}^{\text{th}}$ derivative. We formulate this result as follows:

Theorem 5.7 *Suppose that* $\overline{L}$ *is even,* $k > l$ *holds, and assume that* $E^{(\overline{L}-1)}(0) = 0$ *for all appropriate choices of* τ *etc. which ensure* $E^{(j)}(0) = 0$ *for* $j = 1, \ldots, \overline{L} - 2$. *Then* $E^{(\overline{L})}(0)$ *can be made negative, together with* $E^{(j)}(0) = 0$ *for* $j = 1, \ldots, \overline{L} - 1$.

Proof We essentially proceed in the same way as in the case "$\overline{L}$ odd". We need

Theorem 5.8 (Normal form) *Let* $k > l$ *be satisfied. If* $\overline{L}$ *is even we define* τ *by* (5.24) *and set*

$$\begin{aligned} D_t^{\Gamma}\phi(0) &:= \delta\epsilon^{\Gamma+1} c^{\Gamma+1} i^{\Gamma} w^{-s} + \overline{\delta}\epsilon^{\Gamma+1}(\overline{c})^{\Gamma+1}(-i)^{\Gamma} w^s + \cdots, \\ D_t^{\Gamma+1}\phi(0) &:= \omega\epsilon^{\Gamma+1} c^{\Gamma+1} i^{\Gamma+1} w^{-s} + \overline{\omega}\epsilon^{\Gamma+1}(\overline{c})^{\Gamma+1}(-i)^{\Gamma+1} w^s \\ &\quad + \text{other terms required to remove poles.} \end{aligned} \tag{5.42}$$

As in Theorem 5.5, *we have assumption* (V); *i.e. for* $w = 0$, *the* δ*-linear term of the* $\overline{L}^{\text{th}}$ *derivative vanishes, otherwise it follows automatically that we can make* $E^{\overline{L}}(0) < 0$.

Finally, we choose $\omega = -\gamma\delta(n+1-s)$ *and obtain*:

$$E^{(\overline{L})}(0) = \frac{4(\overline{L}-1)!}{M!(M+1)!} J_{\overline{L}+1} + O(\delta^2\epsilon^{\overline{L}-1}) + O(\epsilon^{\overline{L}}) + O(\omega^2\epsilon^{\overline{L}-1}) \tag{5.43}$$

with $J_{\overline{L}+1}$ *given by* (5.35), *and* $M := (\overline{L}/2) - 1$, $N := (L/2) - 1$.

Moreover, any δ*-linear term in* (5.43) *does not depend on the first complex components of* $\hat{Z}_w(0)$.

Again, the proof will be deferred until later.

Back to the proof of Theorem 5.7. Here we may assume that (5.41) does not hold; we have chosen

$$\omega := -(n+1-s)\gamma\delta,$$

so that we may invoke the Fundamental Computational Principles of Chap. 4.

It follows that the order of the pole of $D_t^{\Gamma+2}\phi(0)$ is $n+s$ and not $n+s+1$, further implying that for $\beta > \Gamma+1$, $D_t^{\beta}\phi(0)$ does not contribute any δ-linear terms to $E^{(\overline{L})}(0)$. The only δ-linear term of order $\epsilon^{\overline{L}-1}$ remaining is

$$\frac{4(\overline{L}-1)!}{M!(M+1)!} J_{\overline{L}+1} = \frac{4(\overline{L}-1)!}{M!(M+1)!} \operatorname{Re} \int_{S^1} w[D_t^{M+1}\hat{Z}(0)] \cdot \zeta \, dw, \tag{5.44}$$

where

$$\zeta = \Bigg(\ldots, \epsilon^{M+1} k_1 R_{\overline{m}} w^{-(n+1-s)} \\ + \epsilon^{L/2+\Gamma} \delta R_m \left[\kappa - \frac{M!}{N!(\Gamma+1)!}(n+1-s)\mu \right] w^{-s} + \cdots \Bigg).$$

Since $n+1 < 2s$, the pole term of largest order in the last complex component of ζ would be w^{-s}, but since (5.41) does not hold, the only leading pole term is

$$\epsilon^{M+1} k_1 R_{\overline{m}} w^{-(n+1-s)}.$$

(Remember that $1+\overline{m} = 1+m+\Gamma(n+1)+s$.) Again, the choice of ω assures that the term of order $\epsilon^{L/2-1+\Gamma}$ in the last complex component of $[D_t^{M+1}\hat{Z}(0)]_w$ is of the form

$$c^{M+1} i^M \epsilon^{L/2-1+\Gamma} \rho w^{n+s}, \quad \rho \neq 0.$$

Thus $E^{(\overline{L})}(0)$ reduces to

$$E^{(\overline{L})}(0) = \frac{4(\overline{L}-1)!}{M!(M+1)!} \operatorname{Re} \int_{S^1} \epsilon^{\overline{L}-1} c^{\overline{L}} i^{\overline{L}-1} \delta k_1 \rho \overline{R}_m \frac{dw}{w} \\ + O(\delta^2 \epsilon^{\overline{L}-1}) + O(w^2 \epsilon^{\overline{L}-1}) + O(\epsilon^{\overline{L}}).$$

This shows that $E^{(\overline{L})}(0)$ can be made negative. □

Remark 5.3 Before going on to prove the normal form theorems, we note that although we cannot prove theorems on non-vanishing of $E^{(j)}(0)$ for $j = \overline{L}-1, \overline{L}, \overline{L}+1$ in case (A) or (C) it remains true that for $\overline{L}$ odd all derivatives of order $j < \overline{L}$ vanish, and for $\overline{L}$ even that all derivatives of order $j < \overline{L}-1$ vanish.

Now we outline the *proofs of the normal form theorems*, i.e. Theorems 5.3, 5.5, 5.6 and 5.8. On a first reading, you may wish to skip this and move directly on to Theorem 5.9. We have to work out

$E^{(\overline{L})}(0)$ and $E^{(\overline{L}+1)}(0)$ for "$\overline{L}$ odd" (Theorems 5.3 and 5.5)

and

$E^{(\overline{L}-1)}(0)$ and $E^{(\overline{L})}(0)$ for "$\overline{L}$ even" (Theorems 5.6 and 5.8).

So as not to repeat the arguments unnecessarily, we do one "lower derivative" and one "higher derivative", namely:

$$E^{(\overline{L})}(0) \quad \text{in case "}\overline{L}\text{ odd"} \quad \text{(Theorem 5.3)}$$

and

$$E^{(\overline{L})}(0) \quad \text{in case "}\overline{L}\text{ even"} \quad \text{(Theorem 5.7).}$$

We begin by writing down a general formula for $E^{(\overline{L})}(0)$ in case that $\overline{L}$ is odd (for a proof, see Sect. 7.2, (7.1)–(7.5)).

We have

$$E^{(\overline{L})}(0) = J_1 + J_2 + J_3$$

where the terms J_1, J_2, J_3 are defined as follows:

Set

$$T^{\alpha,\beta} := w[D_t^\alpha \hat{Z}(0)]_w D_t^\beta \phi(0).$$

Then,

$$\begin{aligned}
J_1 :=\ & 4\,\mathrm{Re} \int_{S^1} [D_t^{\overline{L}-1} \hat{Z}(0)]_w \cdot (w \hat{X}_w \tau)\, dw \\
& + 4 \cdot (\overline{L} - 1)\, \mathrm{Re} \int_{S^1} [D_t^{\overline{L}-2} \hat{Z}(0)]_w f\, dw \\
& + 4 \sum_{\overline{M} > \frac{1}{2}(\overline{L}-1)}^{\overline{L}-3} \frac{(\overline{L}-1)!}{\overline{M}!(\overline{L}-\overline{M}-1)!}\, \mathrm{Re} \int_{S^1} [D_t^{\overline{M}} \hat{Z}(0)]_w \cdot g_{\overline{L}-\overline{M}}\, dw, \qquad (5.45)
\end{aligned}$$

$$f := T^{1,0} + T^{0,1} = w[\hat{Z}_t(0)]_w \tau + w \hat{X}_w \phi_t(0),$$

$$g_\nu := \sum_{\alpha+\beta=\nu} c_{\alpha\beta}^\nu T^{\alpha,\beta} \quad \text{with } c_{\alpha\beta}^\nu := \frac{\nu!}{\alpha!\beta!}, \ \alpha + \beta + \nu = \overline{L} - 1;$$

$$\begin{aligned}
J_2 :=\ & \sum_{\overline{M}=2}^{\frac{1}{2}(\overline{L}-1)} \frac{2(\overline{L}-1)!}{\overline{M}!\overline{M}!}\, \mathrm{Re} \int_{S^1} [D_t^{\overline{M}} \hat{Z}(0)]_w \cdot h_{\overline{M}}\, dw \\
& + 2(\overline{L}-1)(\overline{L}-2)\, \mathrm{Re} \int_{S^1} [\hat{Z}_t(0)]_w \cdot T^{1,\overline{L}-3}\, dw, \qquad (5.46)
\end{aligned}$$

$$h_{\overline{M}} := \sum_{\alpha=0}^{\overline{M}} \psi(\overline{M}, \alpha) \frac{\overline{M}!}{\alpha(\overline{L}-1-\overline{M}-\alpha)!} T^{\alpha, \overline{L}-1-\overline{M}-\alpha},$$

$$\psi(M, \alpha) := 1 \quad \text{for } \alpha = \overline{M}, \qquad \psi(\overline{M}, \alpha) := 2 \quad \text{for } \alpha \neq \overline{M};$$

$$\begin{aligned}
J_3 :=\ & 4(\overline{L}-1)\, \mathrm{Re} \int_{S^1} w \hat{Z}_{tw}(0) \cdot \hat{X}_w D_t^{\overline{L}-2} \phi(0)\, dw \\
& + 2\,\mathrm{Re} \int_{S^1} w \hat{X}_w \cdot \hat{X}_w D_t^{\overline{L}-1} \phi(0)\, dw. \qquad (5.47)
\end{aligned}$$

We have $J_3 = 0$ since $\hat{X}_w \cdot \hat{X}_w = 0$ and $\hat{Z}_{tw}(0) \cdot \hat{X}_w = 0$ on account of formula (2.51) in Sect. 2.1.

We set $D_t^j \phi(0) = 0$ for $0 < j < \Gamma$, and assume that $\tau = \phi(0)$, $D_t^\Gamma \phi(0)$, and $D_t^{\Gamma+1}\phi(0)$ are given by (5.25)–(5.26). As in the non-exceptional case, the proof of the normal forms essentially (but not entirely) consists in counting orders of pole terms. Considering (5.45) we know from the examples of Chap. 2 that $\hat{X}_w \tau$ is always holomorphic, as are $[D_t^\alpha \hat{Z}(0)]_w$, and $\phi_t(0)$ is chosen in such a way that

$$f := w[\hat{Z}_t(0)]_w \tau + w\hat{X}_w \phi_t(0)$$

is holomorphic. Now we turn to

Proposition 5.7 *Suppose $k > l$. Then the δ-linear terms in the integrands of* (5.45) *are holomorphic if τ, $D_t^\Gamma \phi(0)$, $D_t^{\Gamma+1}\phi(0)$ are chosen as in Theorem* 5.4, *and therefore $J_1 = O(\epsilon^{\overline{L}-1}\delta^2)$, whence $E^{(\overline{L})}(0) = J_2 + O(\epsilon^{\overline{L}-1}\delta^2)$.*

Before proceeding with the proof of Proposition 5.7, we shall make some observations, analogs of which are valid for all the normal form theorems.

1. If in Theorem 5.3, we had set $\delta = 0$, and considered only iterations by τ, then in (5.45) no poles ever form on the first complex components (FCC). Thus, for $\beta > 0$, we may take all δ° pole terms of $D_t^\beta \phi(0) = 0$.
2. If $\delta = 0$, the $\overline{L}^{\text{th}}$ derivative contains only terms of order $\epsilon^{\overline{L}}$, which are in general, impossible to calculate.
3. By taking $\delta \neq 0$, as in Theorem 5.3, all δ-linear poles that develop on the A_js' do so only for $j < 2m - 2n + 1$. Since, for these A_j, $A_j = \lambda_j A_1$, all of these poles may be removed as in the non-exceptional case.
4. For $\beta \geq \Gamma$, $D_t^\beta \phi(0)$ are polynomials in δ with (by (5.1)) no constant term.
5. In calculating the $\overline{L}^{\text{th}}$ derivative, we consider only the δ-linear terms.
6. The order of the δ linear poles of $D_t^\beta \phi(0)$ are bounded by $n + 1 + s$, the δ° poles are of order $n + 1$.

We now need a sequence of lemmata.

Lemma 5.3 *Let $k > l$. Then the exponents of the terms in the FCC of $w[D_t^\alpha \hat{Z}(0)]_w$, $\alpha \leq (\overline{L} - 1)/2$ that do not contain δ (or w), are always greater than or equal to $(n + 1)$. If $\alpha \neq (\Gamma + 1)$, the $D_t^\beta \phi(0)$ can be chosen so that exponents of the δ linear terms in the FCC of $w[D_t^\alpha \hat{Z}(0)]_w$ are greater than or equal to $(n+1)$. Most significantly, the order of the δ-linear pole terms of $D_t^\beta \phi(o)$ is bounded by $n + 1 + s$ and δ-linear pole terms formed in the expression*

$$\sum_{\alpha=0}^{\overline{L}-\overline{M}-1} \frac{(\overline{L} - \overline{M} - 1)!}{\alpha!(\overline{L} - \overline{M} - 1 - \alpha)!} w[D_t^\alpha \hat{Z}(0)]_w D_t^\beta \phi(0) \tag{5.48}$$

may be removed. There are no δ° poles.

Proof The statement about β is clearly true for $\beta = 0$, Γ, $\Gamma + 1$, and is true for $\alpha = 0, \ldots, (\Gamma + 1)$. Suppose $\alpha = \Gamma + 2$. Then, since $k > l$, we have an expansion

$$
\begin{aligned}
w[D_t^{\Gamma+1}\hat{Z}(0)]_w\phi = \epsilon^{\Gamma} \cdot (c_1\delta A_1 w^{-s} + c_2\delta A_{n+2}w^{n+1-s} + c_3\delta A_{2n+3}w^{2n+2-s} + \cdots \\
+ c\delta A_{2m-2n+1}w^{(L-2)(n+1)-s} + \cdots, \ldots).
\end{aligned}
\tag{5.49}
$$

Define

$$
D_t^{\Gamma+1}\phi(0) := -\epsilon^{\Gamma} c_1\delta w^{-(n+1+s)} - \epsilon^{\Gamma} c_2\delta\lambda_{n+2}w^{-s} + \epsilon^{\Gamma} c_1c_2\delta\lambda_{n+2}w^{-s} + \cdots.
$$

Then $w[D_t^{\alpha}\hat{Z}(0)]_w$ has an expansion of the form

$$
(c_4 A_{2n+3}w^{2n+2-s} + \cdots, \ldots),
$$

and so our assertion is true for $\alpha = \Gamma + 2$, $\beta = \Gamma$, $\Gamma + 1$.

Now suppose the lemma is true for all α, $\Gamma + 1 < \alpha \leq J - 1$, $\beta \leq J - 2$

$$
\Gamma + 2 \leq \alpha - 1 < (\overline{L} - 1)/2.
$$

We must show it is true for $\alpha = J$.

So consider

$$
w[D_t^{J-1}\hat{Z}(0)]_w\phi + \frac{(J-1)!}{\alpha!(J-1-\alpha)!}\sum_{\alpha=1}^{\tau-2} w[D_t^{\alpha}\hat{Z}(0)]_w D_t^{\beta}\phi(0) + w\hat{Z}_w D_t^{J-1}\phi(0).
\tag{5.50}
$$

By induction, the lowest power of the δ-linear term in the FCC of $w[D_t^{J-1}\hat{Z}(0)]_w$ is $(n+1)$. Thus $w[D_t^{J-1}\hat{Z}(0)]_w\phi$ has no δ-linear pole in the FCC, and thus this term makes no contribution to $D_t^{J-1}\phi(0)$. In

$$
w[D_t^{J-1}\hat{Z}(0)]_w
\tag{5.51}
$$

let us consider the lowest power associated to $A_{2m-2n+1}$, A_l and A_k.

By definition we have that

$$
(l-1) = 2(m+1) - 2(n+1) + \Gamma(n+1) + s.
$$

Since we are at an exceptional branch point we have

$$
m + 1 \geq 2(n+1).
$$

Thus,

$$
(l-1) \geq 1 + m + \Gamma(n+1) + s,
\tag{5.52}
$$

and since $k > l$

$$
(k-1) > 1 + m + \Gamma(n+1) + s.
\tag{5.53}
$$

Therefore the exponent associated to A_l in (5.51) is greater than or equal to

$$
\begin{aligned}
&(l-1) + (1+n) - (L/2 + (\Gamma - 1))(n+1) \\
&\quad \geq 1 + m + (\Gamma + 1)(n+1) + s - (1+m) - (\Gamma - 1)(n+1) = 2n + 2 + s
\end{aligned}
$$

and thus the exponent associated to A_k is also greater than $2(n+1)+s$. We must check the order of the exponent of $A_{2m-2n+1}$. Since $\beta \geq \Gamma, \alpha \leq L/2-1$ $(\alpha+\beta < L/2+\Gamma)$. Thus in $w[D_t^\alpha \hat{Z}(0)]_w$ $A_{2m-2n+1}$ is associated to w^ν, where

$$\nu \geq (L-1)(n+1) - \left(\frac{L}{2}-1\right)(n+1) = \frac{L}{2}(n+1) \geq 2(n+1).$$

Similarly for the other terms of (5.50). Thus, no pole forms on $A_{2m-2n+1}$ and all δ-linear poles are removable. This completes the proof of Lemma 5.3. □

Lemma 5.4 *In* (5.48), *no δ-linear poles or δ° form in the last complex component.*

Proof First consider the term

$$w[D_t^{J-1}\hat{Z}(0)]_w$$

where $J-1 \leq L/2+\Gamma-1$.

Here, we have an expansion of the last complex component of the form

$$c_1 R_{\overline{m}} w^{\nu_1} + c_2 \delta R_m w^{\nu_2}$$

where $\nu_1 \geq 1+\overline{m}-(J-1)(n+1)$. But

$$1+\overline{m} = 1+m+\Gamma(n+1)+s.$$

Therefore

$$\nu_1 \geq 1+m+(\Gamma+1)(n+1)+s - \left(\frac{L}{2}+\Gamma-1\right)(n+1) = 2n+2+s.$$

Furthermore

$$\nu_2 \geq 1+m - \left[\left(\frac{L}{2}+\Gamma-1\right)-(\Gamma+1)\right](n+1)-s = 2(n+1)-s.$$

Therefore in (5.48), the term (5.51) has no δ-linear pole in the last complex component. What about the other terms of (5.48)?

Now $D_t^\beta \phi(0) = 0$ for $0 < \beta < \Gamma$. If $\beta+\alpha < (\overline{L}-1)/2 = L/2+\Gamma$, then in (5.48) consider a non-zero term

$$w[D_t^\alpha \hat{Z}(0)]_w D_t^\beta \phi(0). \tag{5.54}$$

If $\beta \geq \Gamma, \alpha \leq L/2-1$ and if $\beta \geq \Gamma+1, \alpha \leq L/2-2$. If $\beta = \Gamma$, the last complex component of $w[D_t^\alpha \hat{Z}(0)]_w$ has an expansion of the form

$$c_1 R_{\overline{m}} w^{n+1+s} + c_2 R_m w^{n+1} + c_3 \delta w^{2(n+1)-s} \tag{5.55}$$

and if $\beta \geq (\Gamma+1)$ the expansion has the form

$$c_1 R_{\overline{m}} w^{n+1+s} + c_2 R_n w^{2(n+1)} + c_3 \delta w^{3(n+1)-s}.$$

Thus, (5.54) has no pole. □

Lemmas 5.3 and 5.4 now prove Proposition 5.7.

We now proceed with the *proof of Theorem 5.3*. The basic idea is to arrange the $\overline{L}^{\text{th}}$ derivative terms into those with holomorphic integrands and those with poles at $w = 0$. Among the latter we ignore those terms $D_t^j \phi(0)$, $j > \Gamma + 1$, as in the non-exceptional case. We also have:

Lemma 5.5 *All the terms in the value of $\overline{L}^{\text{th}}$ derivative of the form*

$$w[D_t^\alpha \hat{Z}(0)]_w \cdot [D_t^\gamma \hat{Z}(0)]_w D_t^B \phi(0) \tag{5.56}$$

$\gamma, \alpha < (\overline{L} - 1)/2, \Gamma \leq \beta \leq (\overline{L} - 3)/2$ *do not contribute to the value of the $\overline{L}^{\text{th}}$ derivative.*

Proof By Lemma 5.3, the constancy of the order of the δ-linear pole of $D_t^\beta \phi(0)$, $\beta \geq (\Gamma + 1)$ implies that we may ignore all terms of (5.56) where $\beta > (\Gamma + 1)$. First, let us consider the last complex components of $w[D_t^\alpha \hat{Z}(0)]_w$ and $w[D_t^\gamma \hat{Z}(0)]_w$ which take the forms

$$c_1^\alpha R_{\overline{m}} w^{n+1+s} + c_2^\alpha \delta w^{2(n+1)-s} + c_3^\alpha R_m w^{n+1},$$
$$c_1^\gamma R_{\overline{m}} w^{n+1+s} + c_2^\gamma \delta w^{2(n+1)-s} + c_3^\gamma R_m w^{n+1}.$$

Multiplying these out and considering the order of the pole of $D_t^\beta \phi(0)$, the result follows. By Lemma 5.3 if α or γ is less than $\Gamma + 1$ the first complex components of $w[D_t^\alpha \hat{Z}(0)]_w$ are of the form $Cw^{n+1} + C'w^{n+1} + \cdots$,

$$C \cdot A_1 = C \cdot C = 0, \qquad C \cdot C' \neq 0, \qquad C' \cdot A_1 \neq 0.$$

If α or γ is greater than or equal to $(\Gamma + 1)$ the expansion is of the form

$$C_1 w^{n+1} + C' w^{n+1} + C_2 \delta w^{n+1-s} + \cdots,$$
$$C_1 \cdot A_1 = C_1 \cdot C_1 = 0, \qquad C_2 \cdot A_1 = 0, \qquad C' \cdot A_1 \neq 0.$$

This immediately implies that the only possible contribution is of order $O(\epsilon^{\overline{L}-1}\delta^2)$. □

Consider now the following sum in first term on the right-hand side of the definition of J_2 in formula (5.46):

$$\frac{2(\overline{L} - 1)}{M!M!} \operatorname{Re} \int_{S^1} w[D_t^M \hat{Z}(0)]_w$$
$$\cdot \left\{ \sum \frac{M!}{\alpha!\beta!} \psi(m, \alpha)[D_t^\alpha \hat{Z}(0)]_w D_t^\beta \phi(0) \right\} dw, \quad M = \frac{(\overline{L} - 1)}{2}. \tag{5.57}$$

With regard to the contribution of the last complex component to (5.57), the growth estimates of Lemma 5.3 ensure that $D_t^\beta \phi(0)$ for $\beta > \Gamma + 1$ play no role. Thus (5.57) equals

$$\frac{2(\overline{L}-1)!}{M!M!}\operatorname{Re}\int_{S^1} w[D_t^M\hat{Z}(0)]_w\cdot[D_t^M\hat{Z}(0)]_w\tau\,dw$$
$$+\frac{2M!}{(L/2)!\Gamma!}\operatorname{Re}\int_{S^1} w[D_t^{L/2}\hat{Z}(0)]_w\cdot[D_t^M\hat{Z}(0)]_w D_t^\Gamma\phi(0)\,dw$$
$$+\frac{2M!}{(L/2-1)!(\Gamma+1)!}\operatorname{Re}\int_{S^1} w[D_t^N\hat{Z}(0)]_w\cdot[D_t^M\hat{Z}(0)]_w D_t^{\Gamma+1}\phi(0)\,dw,$$
$$N=L/2-1. \tag{5.58}$$

From formula (5.31) we know that the last complex component of

$$w[D_t^M\hat{Z}(0)]_w=i^Mc^M\{\epsilon^Mk_1R_{\overline{m}}w^s+\epsilon^{L/2+\Gamma-1}\kappa\delta R_mw^{n+1-s}\}+\cdots,$$

whereas $w[D_t^{L/2}\hat{Z}(0)]_w$ has a last complex component expansion of the form

$$i^{L/2}c^{L/2}(\epsilon^{L/2}c_1R_{\overline{m}}w^{n+1-s}+\epsilon^{L/2-1}c_2\delta R_mw^{n+1-s}+\cdots). \tag{5.59}$$

Multiplying the expressions, we see that $D_t^\Gamma\phi(0)$ contributes only a term of order $O(\epsilon^{\overline{L}-1}\delta^2)$ and therefore can be ignored.

Thus we are left with only two terms in (5.58), our normal form. It remains to show that the δ-linear terms of the FCC of (5.58) do not contribute to the $\overline{L}^{\text{th}}$ derivative. This follows as in Lemma 5.5 and the proof of Theorem 5.3 is complete.

Now we write down a general formula for $E^{(\overline{L})}(0)$ in case that $\overline{L}$ is even (see Chap. 4, formula (4.66)). We have

$$E^{(\overline{L})}(0)=J_1+J_2+J_3+J_4+J_5 \tag{5.60}$$

with $J_1,\ldots,J_5$ defined as follows:

$$J_1:=4\operatorname{Re}\int_{S^1} w[D_t^{\overline{L}-1}\hat{Z}(0)]_w\cdot\hat{X}_w\tau\,dw;$$
$$J_2:=\sum_{\overline{M}=\overline{s}+1}^{\overline{L}-2}\frac{4(\overline{L}-1)!}{\overline{M}!(\overline{L}-\overline{M}-1)!}\operatorname{Re}\int_{S^1}[D_t^{\overline{M}}\hat{Z}(0)]_w\cdot g_{\overline{L}-\overline{M}-1}\,dw,$$
$$\overline{s}:=\overline{L}/2;$$
$$J_3:=\frac{4(\overline{L}-1)!}{\overline{s}!(\overline{s}-1)!}\operatorname{Re}\int_{S^1}[D_t^{\overline{s}}\hat{Z}(0)]_w\cdot g_{\overline{s}-1}\,dw$$
$$+\frac{2(\overline{L}-1)!}{\sigma!\sigma!}\operatorname{Re}\int_{S^1}[D_t^\sigma\hat{Z}(0)]_w\cdot h_\sigma dw,$$
$$\sigma=\overline{s}-1=\overline{L}/2-1=M,$$
$$h_\sigma:=\sum_{\alpha=0}^{\sigma}c_{\alpha\beta}^\sigma\psi(\sigma,\alpha)w[D_t^\alpha\hat{Z}(0)]_wD_t^\beta\phi(0), \tag{5.61}$$
$$\beta=(\overline{L}-1)-(\overline{s}-1)-\alpha=\overline{L}-\overline{s}-\alpha;$$

$$J_4 := \sum_{\overline{M}=2}^{\overline{s}-1} \frac{2(\overline{L}-1)!}{\overline{M}!\overline{M}!} \operatorname{Re} \int_{S^1} [D_t^{\overline{M}} \hat{Z}(0)]_w \cdot h_{\overline{M}} \, dw$$

$$+ \frac{2(\overline{L}-1)!}{(\overline{L}-3)!} \operatorname{Re} \int_{S^1} w \hat{Z}_{tw}(0) \cdot \hat{Z}_{tw}(0) D_t^{\overline{L}-3} \phi(0) \, dw,$$

$$\alpha + \beta + \overline{M} = \overline{L} - 1;$$

$$J_5 := 4(\overline{L}-1) \operatorname{Re} \int_{S^1} w \hat{Z}_{tw}(0) \cdot \hat{X}_w D_t^{\overline{L}-2} \phi(0) \, dw$$

$$+ 2 \operatorname{Re} \int_{S^1} w \hat{X}_w \cdot \hat{X}_w D_t^{\overline{L}-1} \phi(0) \, dw.$$

We have $J_5 = 0$ because of formula (2.36) in Chap. 2. Moreover, as for the case "$\overline{L}$ odd", we obtain $J_1 = 0$ and $J_2 = O(\delta^2)$ since the δ-linear and w-linear terms in the integrands are holomorphic. Thus we have

$$E^{(L)}(0) = J_3 + J_4 + O(\epsilon^{\overline{L}-1}\delta^2) + O(\epsilon^{\overline{L}-1}\omega^2) + O(\epsilon^{\overline{L}}). \tag{5.62}$$

Now, using the same reasoning as in $\overline{L}$ odd, $J_4 = O(\epsilon^{\overline{L}-1}\delta^2) + O(\epsilon^{\overline{L}-1}\omega^2) + O(\epsilon^{\overline{L}})$. And again, as in the case of $\overline{L}$ odd, in J_3 all terms involving $D_t^\beta \phi(0)$, $\beta > (\Gamma + 1)$ do not contribute δ-linear terms. For $\beta > (\Gamma + 1)$, it follows from our Fundamental Computational Principle that $D_t^{\Gamma+1}\phi(0)$ is used to kill a pole and the only other generator is c/z^{n+1}.

These considerations yield normal form Theorem 5.8, and the same arguments show that it does not depend on the FCC.

Now we summarize the principal results of the preceding discussion in

Theorem 5.9 *Let $\hat{X} \in C^\infty(\overline{B}, \mathbb{R}^3)$ be a nonplanar minimal surface with the exceptional branch point $w = 0$ of order n and index m. Suppose that $\hat{X}$ is given in normal form at $w = 0$, let L be the integer $2(m+1)/(n+1)$, and let $k, l \in \overline{\mathbb{N}}$ be the indices of $\hat{X}$ introduced in Definition 5.1. Then we have:*

(i) *If $k = l = \infty$ then the Taylor expansion of $\hat{X}$ has the form*

$$\hat{X}(w) = X_0 + \operatorname{Re} \sum_{j=1}^{\infty} \tilde{F}_j w^{j(n+1)} \quad \text{for } w \in B \tag{5.63}$$

with $X_0 \in \mathbb{R}^3$ and $\tilde{F}_j \in \mathbb{C}^3$.

(ii) *If not $k = l = \infty$, and $k > l$ is satisfied, then the following holds:*

Given $\epsilon_0 > 0$ and $\mu \in \mathbb{N}_0$, there is a C^∞-diffeomorphism ϕ of ∂B onto itself such that the harmonic extension $\hat{Y} \in C^\infty(\overline{B}, \mathbb{R}^3)$ of $Y := X \circ \phi$ with $X = \hat{X}|_{\partial B}$ satisfies

$$\|\phi - \mathrm{id}_{\partial B}\|_{C^\mu(\partial B, \mathbb{R}^2)} < \epsilon_0, \qquad \|\hat{Y} - \hat{X}\|_{C^\mu(\overline{B}, \mathbb{R}^3)} < \epsilon_0 \tag{5.64}$$

and

$$D(\hat{Y}) < D(\hat{X}).$$

In other words, $\hat{X}$ cannot be a relative minimizer of D "with respect to its own boundary".

Remark 5.4 We first note that the assumption "$\hat{X} \in C^\infty(\overline{B}, \mathbb{R}^3)$" in the previous discussion is only needed when we consider the varied energy functional $E(t)$ produced by variations of the boundary values $X = \hat{X}|_{\partial B}$. The assumption "$\hat{X} \in C^\infty(B, \mathbb{R}^3)$" suffices for the definition of $n, m, L, k, l, \overline{m}$ given at the beginning of this chapter.

Secondly we observe that it is not relevant that the parameter domain of the minimal surface $\hat{X} \in C^\infty(\overline{B}, \mathbb{R}^3)$ in Theorem 5.9 is the unit disk $B_1(0)$. Instead we could take any other disk $B' = B_r(0)$. For instance, if the minimal surface $X : B \to \mathbb{R}^3$ is merely of class $C^0(\overline{B}_1, \mathbb{R}^3)$, we may take $B_r(0)$ with $0 < r < 1$ as domain B to which Theorem 5.9 is applied.

These two observations will be used in the proof of the next result.

In the sequel, Γ will again denote a closed, rectifiable Jordan curve, and $\mathcal{C}(\Gamma)$ is the usual class of disk-type surfaces $\hat{Z} : B \to \mathbb{R}^3$ bounded by Γ.

Theorem 5.10 *Let $\hat{X} \in \mathcal{C}(\Gamma)$ be a nonplanar minimal surface $\hat{X} : B \to \mathbb{R}^3$ in normal form at the exceptional branch point $w = 0$, having the indices $k, l \in \overline{\mathbb{N}}$ as introduced in Definition* 5.1. *Then we have*:

(i) *$k \neq l$ (i.e. $k = l$ cannot happen)*;
(ii) *$\hat{X}$ is not a relative minimizer of A in $\mathcal{C}(\Gamma)$ if $k > l$ is satisfied.*

Proof As $\hat{X} \in C^\infty(B, \mathbb{R}^3)$, the indices $k, l \in \overline{\mathbb{N}}$ are well defined.

(i) Suppose that $k = l = \infty$. Then $\hat{X}(w)$ for $w \in B$ is given by (5.63), according to Theorem 5.9 and Remark 5.4. Set

$$w_r^1 := r \exp\left(\frac{1}{n+1} 2\pi i\right), \quad w_r^2 := r \exp\left(\frac{2}{n+1} 2\pi i\right) \quad \text{for } 0 < r < 1,$$
$$w^1 := \exp\left(\frac{1}{n+1} 2\pi i\right), \quad w^2 := \exp\left(\frac{2}{n+1} 2\pi i\right), \quad \delta_n := |w^1 - w^2| > 0,$$

and note that $w_r^1, w_r^2 \in B$; $w^1, w^2 \in \partial B$; furthermore,

$$|w^1 - w_r^1| = 1 - r, \qquad |w^2 - w_r^2| = 1 - r.$$

Since the boundary values $X = \hat{X}|_{\partial B}$ yield a homeomorphism from ∂B onto Γ, we have

$$\epsilon(\delta_n) := \inf\{|X(w') - X(w'')| : w', w'' \in \partial B \text{ and } |w' - w''| = \delta_n\} > 0.$$

Therefore,

$$\begin{aligned} 0 < \epsilon(\delta_n) &\le |X(w^1) - X(w^2)| \\ &\le |\hat{X}(w_r^1) - \hat{X}(w_r^2)| + |X(w^1) - \hat{X}(w_r^1)| + |X(w^2) - \hat{X}(w_r^2)|. \end{aligned}$$

Since $(w_r^1)^{n+1} = (w_r^2)^{n+1}$, we infer from (5.63) that $\hat{X}(w_r^1) = \hat{X}(w_r^2)$, and so,

$$0 < \epsilon(\delta_n) \le |X(w^1) - \hat{X}(w_r^1)| + |X(w^2) - \hat{X}(w_r^2)| \to 0$$

as $r \to 1$, a contradiction. Consequently, at least one of the indices k, l has to be finite, hence $k \neq l$, and so it makes sense to require that $k > l$ holds on B for $\hat{X}$.

(ii) Fix some $r \in (0, 1)$ and set $B' := B_r(0) \subset\subset B$ and $\hat{X}' := \hat{X}|_{\overline{B'}}$. Then the indices k, l, n, m, L for $\hat{X}' \in C^\infty(\overline{B'}, \mathbb{R}^3)$ are the same as for $\hat{X}$, and so $\hat{X}'$ satisfies $k > l$. By Theorem 5.9 and Remark 5.4 there is a C^∞-diffeomorphism $\phi : \partial B' \to \partial B'$ of $\partial B'$ onto itself such that the harmonic extension $\hat{Z} : \overline{B'} \to \mathbb{R}^3$ of $\hat{X} \circ \phi = \hat{X}' \circ \phi : \partial B' \to \mathbb{R}^3$ to B' satisfies

$$D_{B'}(\hat{Z}) < D_{B'}(\hat{X}'). \tag{5.65}$$

Assume that ϕ is given by

$$\phi(re^{i\theta}) = re^{i\gamma(\theta)}$$

with a 2π-shift periodic function $\gamma \in C^\infty(\mathbb{R})$. Then we define a C^∞-diffeomorphism $\Psi : T \to T$ of the annulus

$$T := \{w \in \mathbb{C} : r \leq |w| \leq 1\}$$

onto itself by setting

$$\Psi(w) := \rho e^{i\gamma(\theta)} \quad \text{for } w = \rho e^{i\theta} \in T,$$

and we note that

$$\Psi(w) = \phi(w) \quad \text{for } w \in \partial B'.$$

Set

$$\hat{X}^*(w) := \begin{cases} \hat{Z}(w) & \text{for } w \in B', \\ \hat{X}(\Psi(w)) & \text{for } w \in T = \overline{B} \setminus B'. \end{cases} \tag{5.66}$$

We have $\hat{Z}(w) = \hat{X}(w)$ for $w \in \partial B'$, and Ψ is a diffeomorphism from ∂B onto itself. Then $\hat{X}^* \in \mathcal{C}(\Gamma) \cap C^0(\overline{B}, \mathbb{R}^3)$, and we have

$$A(\hat{X}^*) = A_{B'}(\hat{Z}) + A_T(\hat{X} \circ \Psi). \tag{5.67}$$

Since $\Psi : T \to T$ is a diffeomorphism and $\hat{X}$ is conformal, we have

$$A_T(\hat{X} \circ \Psi) = A_T(\hat{X}) = D_T(\hat{X}). \tag{5.68}$$

Furthermore, the inequality $A_{B'} \leq D_{B'}$ together with (5.65) yields

$$A_{B'}(\hat{Z}) \leq D_{B'}(\hat{Z}) < D_{B'}(\hat{X}') = D_{B'}(\hat{X}). \tag{5.69}$$

Combining (5.65)–(5.69), we arrive at

$$A(\hat{X}^*) < D_{B'}(\hat{X}) + D_T(\hat{X}) = D(\hat{X}) = A(\hat{X}),$$

taking the conformality of $\hat{X}$ into account. This completes the proof of (ii). □

Corollary 5.1 *Let $\hat{X} \in \mathcal{C}(\Gamma)$ be a nonplanar minimal surface $\hat{X} : B \to \mathbb{R}$ in normal form at $w = 0$ which is a relative minimizer of D. Then $w = 0$ can only be an exceptional branch point of $\hat{X}$ if the indices $k, l \in \overline{\mathbb{N}}$ satisfy $k \neq l$ and $k > l$ does not hold.*

Remark 5.5 On account of this corollary and of Remarks 5.1 and 5.2, Theorem 5.1 will be proved if we can show that $w = 0$ cannot be an exceptional branch point which does not satisfy $k > l$. This will be proved in the next chapter where we derive a new local representation $\hat{Y} := \hat{X} \circ \psi^{-1}$ with possibly new indices k', l', but $n' = n, m' = m, L' = L$, such that $k' = \infty$. Then either $l' = \infty$ or $l' < \infty$. In the latter case we will be able to show that $\hat{X}$ cannot be a minimizer of A in $\mathcal{C}(\Gamma)$. Thus it remains to show that not $k' = l' = \infty$. This cannot be so easily excluded as in Theorem 5.10(i), since the local minimal surface $\hat{Y} : B_{\rho'}(0) \to \mathbb{R}^3$ might not satisfy a Plateau boundary condition. However, $w = 0$ is now a "false branch point" of $\hat{X}$, and this can be excluded by the reasoning of Gulliver/Osserman/Royden [1]. We give our own brief proofs of this result in the care of smooth boundaries in Chap. 7.

Chapter 6
Exceptional Branch Points Without the Condition $k > l$

Now let $\hat{X} \in \mathcal{C}(\Gamma)$ be a nonplanar minimal surface in normal form at the exceptional branch point $w = 0$ with $k \neq l$ which does not satisfy $k > l$. We have

$$\hat{X}_w(w) = (A_1 w^n + A_2 w^{n+1} + \cdots, R_m w^m + \cdots)$$

with $m > n$, $A_1 \neq 0$, $R_m \neq 0$. As in Chap. 5 we introduce C_j and $C'_j \in \mathbb{R}^2$ by

$$\begin{aligned} C_j &:= A_j, \quad C'_j := 0 \quad \text{if } A_j \cdot A_1 = 0, \\ C_j &:= 0, \quad C'_j := A_j \quad \text{if } A_j \cdot A_1 \neq 0. \end{aligned}$$

Set

$$\Theta_j := (j-1) + (n+1);$$

then

$$w\hat{X}_w(w) = \left(A_1 w^{n+1} + \sum_{j>1} C_j w^{\Theta_j} + \sum_{j>1} C'_j w^{\Theta_j}, R_m w^{m+1} + \cdots \right).$$

Lemma 6.1 *We have* $C'_j = \lambda'_j A_1 + \nu'_j \overline{A}_1$.

Proof A_1 and $\overline{A}_1$ are linearly independent since $A_1 \cdot A_1 = 0$, $A_1 \cdot \overline{A}_1 \neq 0$, and $\dim_{\mathbb{C}} \mathbb{C}^2 = 2$. □

Write $C_j = \lambda_j A_1$. Then we have

$$w\hat{X}_w(w) = \left(A_1 \left(\sum_j \lambda_j w^{\Theta_j} + \sum_j \lambda'_j w^{\Theta_j} \right) + \overline{A}_1 \sum_j \nu'_j w^{\Theta_j}, R_m w^{m+1} + \cdots \right).$$

Let us introduce the holomorphic function $g(w)$ by

$$g(w) = \frac{w^{n+1}}{n+1} + \sum_{j>1} [(\lambda_j + \lambda'_j)/\Theta_j] w^{\Theta_j}. \tag{6.1}$$

A. Tromba, *A Theory of Branched Minimal Surfaces*,
Springer Monographs in Mathematics,
DOI 10.1007/978-3-642-25620-2_6,

So we can write

$$g(w) = w^{n+1}\psi_1(w)$$

where $\psi_1(w)$ is a holomorphic function on B which can be written as

$$\psi_1(w) = \frac{1}{n+1} + \psi_2(w),$$

where $\psi_2(w)$ is a holomorphic function on B of the form

$$\psi_2(w) = a_1 w + a_2 w^2 + \cdots.$$

Therefore, close to $w = 0$, we can extract the $(n+1)^{\text{th}}$ root of $\psi_1(w)$, i.e. there is a neighbourhood $B_\rho(0)$ of $w = 0$, $0 < \rho < 1$, and a holomorphic mapping $\varphi : B_\rho(0) \to \mathbb{C}$ with $\varphi(0) = [1/(n+1)]^{\frac{1}{n+1}} > 0$ such that

$$\varphi^{n+1}(w) = \psi_1(w) \quad \text{for } w \in B_\rho(0).$$

Introduce the holomorphic function $\psi : B_\rho(0) \to \mathbb{C}$ by

$$\psi(w) := w\varphi(w) \quad \text{for } w \in B_\rho(0). \tag{6.2}$$

We have $\psi(0) = 0$ and $\psi'(0) > 0$ because of

$$\psi'(w) = \varphi(w) + w\varphi'(w).$$

Thus the mapping $w \mapsto \psi(w)$ is biholomorphic on $B_\rho(0)$ for $0 < \rho \ll 1$, and its inverse ψ^{-1} is a well-defined holomorphic map on $B_{\rho'(0)}$, $0 < \rho' \ll 1$, with $\psi(0) = 0$. We obtain the Taylor expansion

$$\psi^{-1}(z) = c_1 z + \cdots \quad \text{with } c_1 \neq 0,$$

and from $z = \psi(w)$ it follows

$$z^{n+1} = w^{n+1}\varphi^{n+1}(w) = w^{n+1}\psi_1(w) = g(w) \quad \text{for } |w| < \rho.$$

Using the transformation $w = \psi^{-1}(z)$ on $B_{\rho'(0)}$, we can introduce a new minimal surface $Y : B_{\rho'}(0) \to \mathbb{R}^3$ by setting

$$\hat{Y} := \hat{X} \circ \psi^{-1}. \tag{6.3}$$

Then we obtain the local expansion

$$\hat{Y}(z) = X_0 + \text{Re}(A_1 z^{n+1} + \overline{A}_1 \gamma(z), R'_m z^{m+1} + \cdots) \tag{6.4}$$

with $A_1 \neq 0$, $R'_m \neq 0$, and $\gamma(z) := a z^{2m-n+1} + \cdots$.

Thus we arrive at

Proposition 6.1 *There is a biholomorphic mapping ψ from a neighbourhood of $w = 0$, satisfying $\psi(0) = 0$, such that $\hat{Y} = X \circ \psi^{-1} : B_{\rho'}(0) \to \mathbb{R}^3$ with $0 < \rho' \ll 1$ is a mapping in normal form at the exceptional branch point $w = 0$ which satisfies $k' = \infty$.*

Now we are going to verify

Proposition 6.2 *Consider the minimal surface* $\hat{Y} = \hat{X} \circ \psi^{-1} : B_{\rho'}(0) \to \mathbb{R}^3$, $0 < \rho' \ll 1$, *from Proposition* 6.1 *with the new indices* $k', l' \in \overline{\mathbb{N}}$. *Then we have: Either* $k' = l' = \infty$, *or else there is a mapping* $\hat{X}^* \in \mathcal{C}(\Gamma)$ *such that* $A(\hat{X}^*) < A(\hat{X})$.

Proof Choose r_1 and r_2 with $0 < r_1 < r_2 < \rho'$ and set

$$B' := B_{r_1}(0), \qquad B'' := B_{r_2}(0), \qquad T := \overline{B''} \setminus B' = \{z \in \mathbb{C} : r_1 \le |z| \le r_2\}.$$

Moreover, we apply Theorem 5.9 and Remark 5.4 of Chap. 5 to Y, assuming that we have not $k' = l' = \infty$. Then, for any $\epsilon_0 > 0$, there is a C^∞-diffeomorphism ϕ of $\partial B'$ onto itself such that

$$\|\phi - \mathrm{id}_{\partial B'}\|_{C^1(\partial B'', \mathbb{R}^2)} < \epsilon_0, \tag{6.5}$$

and that the harmonic extension $\hat{Z} \in C^\infty(\overline{B'}, \mathbb{R}^3)$ of $Z := Y \circ \phi$ with $Y := \hat{Y}|_{\partial B'}$ satisfies

$$D_{B'}(\hat{Z}) < D_{B'}(\hat{Y}). \tag{6.6}$$

Let ϕ be represented by

$$\phi(re^{i\theta}) = re^{i\gamma(\theta)} \tag{6.7}$$

with a 2π-shift periodic function $\gamma \in C^\infty(\mathbb{R})$. Because of (6.5) we have

$$|\gamma'(\theta) - 1| < \delta(\epsilon_0) < \frac{1}{4} \quad \text{with} \quad \lim_{\epsilon_0 \to +0} \delta(\epsilon_0) = 0. \tag{6.8}$$

Now we define a mapping $\Psi : T \to T$ of the annulus T onto itself by setting

$$\Psi(z) := re^{i\tilde{\gamma}(r,\theta)} \quad \text{for } z = re^{i\theta} \text{ with } r_1 \le r \le r_2,$$
$$\tilde{\gamma}(r,\theta) := [1 - \mu(r)]\gamma(\theta) + \mu(r)\theta, \quad \mu(r) := \frac{r - r_1}{r_2 - r_1}.$$

We note that, for any $r \in [r_1, r_2]$, Ψ provides a one-to-one mapping of the circle $\partial B_r(0)$ onto itself; thus Ψ yields a one-to-one mapping from T onto itself. Furthermore,

$$\Psi_r(re^{i\theta}) = e^{i\tilde{\gamma}(r,\theta)} \left\{ 1 + ir\frac{r_1}{r_2 - r_1}[\gamma(\theta) - \theta] \right\},$$
$$\Psi_\theta(re^{i\theta}) = ir\tilde{\gamma}_\theta(r,\theta)e^{i\tilde{\gamma}(r,\theta)}.$$

Fix some sufficiently small $\epsilon_0 > 0$. Then the Jacobian J_Ψ of Ψ is positive on T, taking (6.8) into account, and therefore Ψ is a diffeomorphism of T onto itself which satisfies

$$\Psi(z) = \phi(z) \quad \text{for } |z| = r_1, \qquad \Psi(z) = z \quad \text{for } |z| = r_2.$$

Define $\hat{Y}^* : B'' \to \mathbb{R}^3$ by

$$\hat{Y}^*(z) := \begin{cases} \hat{Z}(z) & \text{for } z \in B', \\ \hat{Y}(\Psi(z)) & \text{for } z \in T. \end{cases} \tag{6.9}$$

Then $\hat{Y}^* \in H_2^1(B'', \mathbb{R}^3) \cap C^0(\overline{B''}, \mathbb{R}^3)$, $\hat{Y}^*(z) = \hat{Y}(z)$ for $z \in \partial B''$, and

$$A_{B''}(\hat{Y}^*) = A_{B'}(\hat{Z}) + A_T(\hat{Y} \circ \Psi). \tag{6.10}$$

From (6.6) and the conformality of $\hat{Y}$ we infer

$$A_{B'}(\hat{Z}) \leq D_{B'}(\hat{Z}) < D_{B'}(\hat{Y}) = A_{B'}(\hat{Y}), \tag{6.11}$$

and we also have

$$A_T(\hat{Y} \circ \Psi) = A_T(\hat{Y}) \tag{6.12}$$

since Ψ is a diffeomorphism from T onto itself. By virtue of (6.10)–(6.12) we arrive at

$$A_{B''}(\hat{Y}^*) < A_{B'}(\hat{Y}) + A_T(\hat{Y}) = A_{B''}(\hat{Y}). \tag{6.13}$$

Recall that $\hat{Y} = \hat{X} \circ \psi^{-1}$ for the biholomorphic mapping $\psi^{-1} : B_{\rho'}(0) \to B$ and set $U := \psi^{-1}(B'') \subset\subset B$. Then

$$\hat{Y} \circ \psi|_{\overline{U}} = \hat{X}|_{\overline{U}} \quad \text{and} \quad \hat{Y}^* \circ \psi|_{\partial U} = \hat{X}|_{\partial U}$$

where ∂U is regular and real analytic. It follows that

$$\hat{X}^*(w) := \begin{cases} \hat{Y}^*(\psi(w)) & \text{for } w \in \overline{U} \\ \hat{X}(w) & \text{for } w \in \overline{B} \setminus U \end{cases} \tag{6.14}$$

defines a surface $X^* \in \mathcal{C}(\Gamma)$, and we obtain from (6.13) and (6.14)

$$\begin{aligned} A(\hat{X}^*) &= A_U(\hat{Y}^* \circ \psi) + A_{B\setminus U}(\hat{X}) \\ &= A_{B''}(\hat{Y}^*) + A_{B\setminus U}(\hat{X}) \\ &< A_{B''}(\hat{Y}) + A_{B\setminus U}(\hat{X}) \\ &= A_{B''}(\hat{X} \circ \psi^{-1}) + A_{B\setminus U}(\hat{X}) \\ &= A_U(\hat{X}) + A_{B\setminus U}(\hat{X}) = A(\hat{X}). \end{aligned}$$

This completes the proof of Proposition 6.2. □

Suppose now that in Proposition 6.2 we have the case $k' = l' = \infty$. This means there is a biholomorphic mapping $\psi : B_\rho(w_0) \to \mathbb{C}$, $0 < \rho \ll 1$, with $\psi(0) = 0$ such that $\hat{Y}(z) := \hat{X}(\psi^{-1}(z))$ satisfies

$$\hat{Y}(z) = Y_0 + \operatorname{Re} \sum_{j=1}^{\infty} \tilde{F}_j z^{j(n+1)} \quad \text{for } z \in B_{\rho'}(0),\ 0 < \rho' \ll 1,$$

where $\tilde{F}_1 \neq 0$.

This means that $w = 0$ is a "locally analytically false" branch point of $\hat{X}$, and in particular, $w = 0$ *is a false branch point of* $\hat{X}$. However, this is impossible for $\hat{X} \in \mathcal{C}(\Gamma)$, according to the results of Gulliver/Osserman/Royden [1] which we might use here (cf. Sect. 6 of [1], in particular Theorem 6.16). Thus we obtain

Proposition 6.3 *For $\hat{X} \in \mathcal{C}(\Gamma)$ and the mapping $\hat{Y} = \hat{X} \circ \psi^{-1}$ from Proposition* 6.1, *we do not have $k' = l' = \infty$.*

From Propositions 6.1–6.3 we infer:

Theorem 6.1 *Let $\hat{X} \in \mathcal{C}(\Gamma)$ be a nonplanar minimal surface $\hat{X} : B \to \mathbb{R}^3$ in normal form at the exceptional branch point $w = 0$, and suppose that the condition $k > l$ does not hold. Then $\hat{X}$ is not an absolute minimizer of A in $\mathcal{C}(\Gamma)$.*

Combining this theorem with results in Chap. 5, namely Remarks 5.1 and 5.2 and Theorem 5.10, the Second Main Theorem (= Theorem 5.1 of Chap. 5) is proved.

Chapter 7
New Brief Proofs of the Gulliver–Osserman–Royden Theorem

7.1 The First Proof

We would like to present very much simplified proofs of versions of the Gulliver–Osserman–Royden (GOR) theorem [1], in the case Γ is $C^{2,\alpha}$ smooth. In the first proof instead of employing a topological theory of ramified coverings used in (GOR), we introduce a new analytical method of root curves. The surprising aspect of this proof is that it connects the issue of the existence of analytical false interior branch points with boundary branch points. We should note that this fact was also observed by F. Tomi [1] who has found his own very brief proof of (GOR) in the case $\Gamma \in C^{2,\alpha}$ which we also include. We first state our main

Theorem 7.1 *If $\hat{X} \in C(\Gamma)$, $\Gamma \in C^{2,\alpha}$, is a C^0 relative minimum of area then $w = 0$ cannot be an analytically false branch point, and thus $\hat{X}$ cannot have $w = 0$ as a branch point.*

We begin by defining the set:

$$\mathcal{R} = \{p \in S^1 \mid \hat{X}(p) = \hat{X}(q),\ q \in B^0\}.$$

Lemma 7.1 *If $\hat{X} \in \mathcal{C}(\Gamma), \mathcal{R} \neq S^1$.*

Proof Suppose $\mathcal{R} = S^1$. Consider a linear function $\mathcal{L} = \mathbb{R}^3 \to R$ given by $\mathcal{L}(x) = \Sigma a_i x_i$. Then for a fixed $a = (a_1, a_2, a_3)$, the planes are parallel and for b large positive these planes do not intersect $\hat{X}(\overline{B})$. Decrease b until there is a first point of intersection; i.e. $L(\hat{X}(p)) + b = 0$ for some $p \in \overline{B}$. Then for all $p \in B$, $L(\hat{X}(p)) + b \leq 0$. Noting that $\kappa(p) := L(\hat{X}(p)) + b$ is a harmonic function, then by the strong maximum principle for harmonic functions, κ cannot have an interior maximum. Therefore $p \in S^1$ and $L(\hat{X}(g)) + b < 0$ for all $q \in B^0$. But $L(\hat{X}(p)) + b = L(\hat{X}(p')) + b$ for some $p' \in B^0$, a contradiction, proving Lemma 7.1. □

From what we have proved already, we know that if $\hat{X} \in \mathcal{C}(\Gamma)$ is a C^0-relative minimum with an exceptional branch point at $w = 0$, then near $w = 0$ $\hat{X}(w) =$

A. Tromba, *A Theory of Branched Minimal Surfaces*,
Springer Monographs in Mathematics,
DOI 10.1007/978-3-642-25620-2_7,

$Y(g(w))$ for some locally defined (near $w = 0$) embedded minimal surface Y. Moreover

$$\hat{X}_w = (A_1 g'(w) + \overline{A}_1 \rho'(w), R_m \gamma'(w)). \tag{7.1}$$

We now have:

Lemma 7.2 *If $\Gamma \in C^{2,\alpha}(\overline{B})$, then g, ρ, γ are in $C^{2,\alpha}(B)$. If Γ is real analytic then g, ρ, γ are real analytic.*

Proof If $\Gamma \in C^{2,\alpha}$, then $\hat{X} \in C^{2,\alpha}(\overline{B})$ and if Γ is real analytic $\hat{X}$ has a real analytic extension to a neighbourhood of $\overline{B}$. Since $\overline{A}_1 \cdot \overline{A}_1 = 0$, we have

$$(\overline{A}_1 \cdot A_1) g'(w) \in C^{1,\alpha}(\overline{B}).$$

Similarly for ρ and γ. The analytic case follows analogously. □

We begin the proof of Theorem 7.1 by first ruling out analytically false interior branch points in the case $\hat{X}$ has no boundary branch points. This case is quite easy to prove and gives some geometric understanding as to why Theorem 7.1 is true.

Proposition 7.1 *If $\hat{X} \in \mathcal{C}(\Gamma)$ is a C^0 relative minimum of area with an analytically false branch point at $w = 0$, as above, and if $g' \neq 0$ on S^1, then $\mathcal{R} = S^1$.*

By Lemma 7.1 we know that $\mathcal{R} = S^1$ is impossible, thus we obtain:

Theorem 7.2 *If $\hat{X} \in C(\Gamma)$, $\Gamma \in C^{2,\alpha}$, $\hat{X}$ a C^0 relative minimum of area where $g \neq 0$ and $g' \neq 0$ on S^1, then $\hat{X}$ cannot have $\omega = 0$ as an analytically false branch point, and thus a C^0 relative minimum of area cannot be branched at all.*

Lemma 7.3 *Under the hypothesis of Theorem 7.2, $\mathcal{R} \neq \emptyset$.*

To facilitate the proof of Lemma 7.3, we introduce:

$\mathcal{A} := \{\sigma : [0, 1] \to \mathbb{C} \mid \sigma$ analytic, embedded, $\sigma(1) \in U$, where U is the unbounded component of $\mathbb{C} \setminus g(S^1)$, $\sigma(0) = 0$, σ avoids the image of the zeros of g' (see remark below) other than 0 and σ is transverse to $g(S^1)\}$.

Remark 7.1 The interior zeros g' are countable and since g is holomorphic on B and C^2 on S^1, the zeros of g' are nowhere dense and this image has Hausdorff 1-dimension zero, Jiang [1]. Thus, Tomi–Tromba [1], the image of these zeros does not disconnect $\mathbb{R}^2$.

Then, for all $\sigma \in \mathcal{A}$, $g^{-1}(\sigma[0, 1])$ is a one-dimensional submanifold of $\overline{B}$ transverse to S^1. We now construct what we call the *root curves* of $\sigma \in a$.

Lemma 7.4 *The equation $g(w) = \sigma(t)$ generates $(n + 1)$ analytic root curves, $\sigma_j(t), 1 \le j \le (n+1)$, each analytic on $0 < t < 1$ and satisfying*

$$\begin{aligned} &\sigma_j(0) = 0, \qquad g(\sigma_j(1)) = u \\ &g(\sigma_j(t)) = \sigma(t) \quad \forall j \\ &\rho(\sigma_j(t)) = \rho(\sigma_i(t)) \quad \forall i, j \\ &\gamma(\sigma_j(t)) = \gamma(\sigma_i(t)) \quad \forall i, j \\ &g(\sigma_j(t)) = g(\sigma_i(t)) \quad \forall i, j \end{aligned} \tag{7.2}$$

for all t such that $\sigma_i(t) \in \overline{B}$.

Proof Along $\sigma(t)$, we may define the $(n + 1)^{\text{th}}$ roots of $\sigma(t)$, namely $\xi_1(t), \dots, \xi_{n+1}(t)$, $\xi_i(t)^{n+1} = \sigma(t)$. About a neighbourhood of 0, and since $w \mapsto w\varphi(w)$ is a local diffeomorphism, the $\{\sigma_j\}$ are the $(n + 1)$ solutions to the equation

$$w\varphi(w) = \xi_i(t). \tag{7.3}$$

For $t > 0$, $g^{-1}(\sigma(t))$ is a one-dimensional manifold and the $\{\sigma_i(t)\}$ constitute $(n + 1)$ components of $g^{-1}(\sigma(t))$. By constructing g maps each $\sigma_i(t)$ diffeomorphically onto its image, and near $w = 0$

$$\hat{X}(w) = Y(g(w)), \qquad [\omega\varphi(\omega)]^{n+1} = g(\omega) \tag{7.4}$$

for some Y. This implies that, near $w = 0$, $\rho(w) = \tilde{\rho}(g(w)), \gamma(w) = \tilde{\gamma}(g(w))$ and from this (7.2) follows by analyticity. □

From (7.2) we immediately have:

Lemma 7.5

$$\hat{X}(\sigma_j(t)) = \hat{X}(\sigma_i(t)) \tag{7.5}$$

for all i, j such that $\sigma_i(t), \sigma_j(t) \in \overline{B}$.

Lemma 7.6 *For each index j, there is a $t_j, 0 < t_j \le 1$ such that $\sigma_j(t_j) \in S^1$.*

Proof If not $\sigma_j(t) \in B^0$ for all t, hence $\sigma_j(1) \in B^0$. Thus $g(\sigma_i(1)) = \sigma(1) \in U$ and g maps an interior point of B to a point in the unbounded component of $\mathbb{C} \setminus g(S^1)$, a clear impossibility. □

Thus there must be an index τ and a first time t_τ such that $\sigma_\tau(t_\tau) \in S^1$. By (7.5) and monotonicity $\sigma_j(t_\tau) \in B^0$, for all $j \ne \tau$. Thus $\sigma_\tau(t_\tau) \in \mathcal{R}$ and $\mathcal{R} \ne \emptyset$, proving Lemma 7.4.

The next two lemmas conclude the proofs of Proposition 7.1 and Theorem 7.2.

Lemma 7.7 *$\mathcal{R}$ is open and non-empty.*

Lemma 7.8 *Under the hypotheses of Theorem 7.2 $\mathcal{R}$ is closed.*

The basic idea is to show that each $\sigma \in \mathcal{A}$ generates a point of $\mathcal{R}$. To achieve this we construct a map

$$\mathcal{A} \xrightarrow{\Phi} \mathcal{R} \subset S^1.$$

For each $\sigma \in a$, there is an index τ, a root curve σ_τ and a first time t_τ such that t_τ is the first time σ_τ hits S^1, and for all indices $j \neq \tau, \sigma_j(t_\tau) \in B^0$, for if $\sigma_j(t_\tau) \in S^1$, (7.5) would violate boundary monotonicity of $\hat{X}$. Now t_τ depends smoothly on σ, i.e. $\sigma \to t_\tau(\sigma)$ is smooth. By smooth we mean the following: If $h : [0,1] \to \mathbb{C}$, $h(0) = 0$ is analytic and small in any C^r norm, $r \geq 3$, then $(\sigma + h) \in \mathcal{A}$. Then

$$s \mapsto t_\tau(\sigma + sh)$$

is C^1. We define the map

$$\Phi : \mathcal{A} \to \mathcal{R}$$

by

$$\Phi(\sigma) = \sigma_\tau(t_\tau) \in S^1.$$

Proof of Lemma 7.7 Let $\sigma \in \mathcal{A}$ be fixed, $p = \sigma_\tau(t_\tau)$. Furthermore, let (π, w) be a normal bundle of S^1 such that the derivative of the projection map $D\pi(p) : \mathbb{R}^2 \to T_pS^1$, has the property that

$$\sigma_\tau'(t_\tau) \in \operatorname{Ker} D\pi(p).$$

Given an analytic map h, define $h_\tau(t)$ by

$$Dg(p)h_\tau(t) = h(t). \tag{7.6}$$

We have $\Phi(\sigma) = \sigma_\tau(t_\tau(\sigma))$, thus

$$D\Phi(\sigma)h = h_\tau(t_\tau(\sigma)) + \sigma_\tau'(t_\tau(\sigma))Dt_\tau(\sigma)h.$$

Since $\pi(\sigma_\tau(\sigma)) = \sigma_\tau(t_\tau(\sigma))$,

$$D\pi \cdot D\Phi(\sigma)h = D\Phi(\sigma)h = D\pi h_\tau(t_\tau(\sigma)).$$

If $v \in T_pS^1$ is arbitrary, pick h so that

$$h_\tau(t_\tau(\sigma)) = v.$$

Then $D\Phi(\sigma)h = v$ and $D\Phi$ is surjective implying that the range of Φ is open, and Lemma 7.7 is proved. □

Proof of Lemma 7.8 Assume we have a sequence $\sigma_{\tau_m}^m(t_{\tau_m}) \to p \in S^1$. We claim that all other root curves must be bounded away from S^1. Clearly $\{\sigma_j^m(t_{\tau_m})\}$, $j \neq \tau_m$ cannot have an accumulative point on S^1 other than p, since by (7.5) this would violate monotonicity. Also p cannot be an accumulation point since g is a local diffeomorphism and (7.2) would be violated.

By picking a subsequence, we may assume that $\sigma_i^m(t_{\tau_m}), i \neq \tau_m$ converge to $p_1, \ldots, p_n \in B^0$, n the order of the branch point $\omega = 0$. Let $p_0 = p$. By changing variables we may assume that around each $p_\ell, \ell \neq 0$

$$g(\omega) = a_\ell(\omega - p_\ell)^{r_\ell} + g(p).$$

Let V be a neighbourhood of $g(p)$ and W_0 of p with $g : W_0 \to V$ a diffeomorphism and $W_\ell, \ell \neq 0$ neighbourhoods of p_ℓ such that $W_\ell \subset g^{-1}(V)$, and $W_\ell \cap S^1 \neq \mathbb{C}$. Let N be large enough so that $\sigma^N_{\tau_N}(t_{\tau_N}) \in W_0$, and if $j \neq \tau_N$, $\sigma^N_j(t_{\tau_N}) \in W_\ell$ for some ℓ. Let $\tau = \tau_N, \sigma_\tau := \sigma^N_{\tau_N}$. Let $\tilde{\sigma}_\tau$ be a C^∞ embedded mapping agreeing with σ_τ outside W_0, $\tilde{\sigma}_\tau(t_\tau) = p$ and the image of $\tilde{\sigma}_\tau$ avoiding the pre-image of the image of the interior zeros of g'. By approximation, we may assume that $\tilde{\sigma}_\tau$ is analytic. Let $\tilde{\sigma} := g(\tilde{\sigma}_\tau)$. Then $p = \tilde{\sigma}_\tau(t_\tau) = \Phi(\tilde{\sigma}) \in \mathcal{R}$, and so $\mathcal{R}$ is closed. This completes the proof of Lemma 7.8 and Theorem 7.2. □

We now want to prove Theorem 7.1 by finding a way to apply Theorem 7.2 to a disk of radius slightly less than 1, even though $\hat{X}$ restricted to the boundary of this disk need not be monotonic. Thus monotonicity must be replaced by other conditions. This is the content of Lemmas 7.10 and 7.11. Let $p_1, \ldots, p_\ell$ be the finite number of branch points of $\hat{X}$ on S^1, and $W_1, \ldots, W_\ell$ neighbourhoods so that the boundary expansion

$$\hat{X}_W(w) = (w - p_j)^{r_i} F_j(w). \tag{7.7}$$

$F_j(p_j) \neq 0$, holds. Let $K \subset S^1$, $K := \sim(\cup W_j) \cap S^1$, $\sim$ meaning "complement of" and define

$$\mathcal{A}_K := \{\sigma \in \mathcal{A} \mid \sigma_\tau(t_\tau) \in K\}. \tag{7.8}$$

We now have:

Lemma 7.9 *For $j \neq \tau$ define*

$$I_j(K) := \inf_{\sigma \in a_K} \{\mathrm{dist}(\sigma_j(t_\tau), S^1)\} \tag{7.9}$$

where dist := *distance.*

Then, for all $j \in \tau$, $I_j(K) > 0$.

Proof Suppose the contrary. Let $\{\sigma^m\} \in Q$ be a sequence with $\mathrm{dist}(\sigma^m_j(t_{\tau_m}), S^1) \to 0$. By passing to a subsequence we may assume that $\sigma^m_{\tau_m}(t_{\tau_m}) \to p \in S^1$ and $\sigma^m_{j_m}(t_{\tau_m}) \to q \in S^1$. By monotonicity and (7.5), $p = q$. Since p is not a branch point either g, ρ, γ (cf. (7.1)) must have a non-vanishing derivative at p, say $\rho'(p) \neq 0$. Then ρ is locally one to one, but $\rho(\sigma^m_{i_m}(t_{\tau_m})) = \rho(\sigma^m_{\tau_m}(t_{\tau_m}))$, implying that for m large $\sigma^m_{j_m}(t_{\tau_m}) = \sigma^m_{\tau_m}(t_{\tau_m})$, a contradiction. □

Now let $W = W_k$ for some k, $\mathcal{A}_{W \cap S^1}$ defined as was $\mathcal{A}_K$, and $I_j(W \cap S^1)$ defined as was $I_j(K)$ (cf. (7.9)) (i.e. $W \cap S^1$ replaces K).

Lemma 7.10 *For W sufficiently small, and for all $j \neq \tau$*

$$I_j(W \cap S^1) > 0.$$

Since the proof of Lemma 7.10 is a bit technical, let us first complete the proof of Theorem 7.1, assuming Lemmas 7.9 and 7.10.

Proof of Theorem 7.1 Let $\{r_m\}$ be a sequence of radii converging to 1 from below with $g|S_{r_m} \neq 0$ and $g'|S_{r_m} \neq 0$, S_{r_m} the circle of radius r_m. It follows from Lemmas 7.9 and 7.10, that for m sufficiently large, no two root curves can simultaneously hit S_{r_m}. Let $\mathcal{R}_m := \{p \in \mathcal{R}_m | \hat{X}(p) = \hat{X}(q)$, where q is in the interior of the disk of radius $r_m\}$. Then Theorem 7.2 and previous lemmas can be applied to this situation to conclude that $\mathcal{R}_m = S_{r_m}$ and thus $w = 0$ cannot be a branch point, thus proving Theorem 7.1. □

Proof of Lemma 7.10 Assume that on $\overline{W}$, $q = q_k$,

$$\rho(w) = (w - q)^r \tilde{\rho}(w) + c \tag{7.10}$$

$\tilde{\rho}(q) \neq 0$, r odd, $\tilde{\rho} \in C^1$. Identify q with 0 in the Poincaré upper half-plane $\mathcal{H}$. On $\mathcal{H}$, the mapping $z \to z^r$ divides any disk centred at 0 into r disjoint open sets on which $z \to z^r$ is one to one.

Since near 0, $z \mapsto z^r \tilde{\rho}(z)$ can be written as $w(z)^r$, $z \to w(z)$ a local diffeomorphism, we immediately obtain:

Lemma 7.11 *There are $\nu \geq 3$ pairwise disjoint open sets $\Omega_1, \ldots, \Omega_\nu$, with $\overline{W} = \cup \overline{\Omega}_i$, for W sufficiently small, and ρ is one to one on each Ω_j.*

We now prove Lemma 7.10. Assume W has been chosen so that Lemma 7.11 applies. Choose a neighbourhood V of c so that a component of $\rho^{-1}(V) \subset W$. Now suppose $I_j(W \cap S^1) = 0$, and let $\sigma^m \in \mathcal{A}_{W \cap S^1}$ with $\sigma^m_{\tau_m}(t^m_{\tau_m}) = p_m \in W$ and $\text{dist}(\sigma^m_j(\tau^m_{t_m}), S^1) \to 0$. By passing to a subsequence we may assume $p_m \to p$. Then $p = q$, otherwise the argument in Lemma 7.9 would yield a contradiction. Thus, we may assume that for some path σ we have:

(i) $\sigma_\tau(t_\tau) \in W \cap S^1$.
(ii) For some $j \neq \tau, \sigma_j(t_\tau) \in \Omega_2 \cap \rho^{-1}(V)$.
(iii) For some s_0, $\sigma_\tau(s) \in \Omega_1 \cap \rho^{-1}(V)$ for all $s \geq s_0$, and $\sigma_j(s) \in \Omega_2 \cap \rho^{-1}(V)$.

To complete the proof of Lemma 7.10, we argue as before. Let $\tilde{\sigma}_\tau$ be a C^∞ path such that

(iv) $\tilde{\sigma}_\tau(0), \tilde{\sigma}_\tau(t_\tau) = q$.
(v) $\tilde{\sigma}_\tau(t) = \sigma_\tau(t), \forall s \leq s_0$.
(vi) $\tilde{\sigma}_\tau[s_0, t_\tau] \in \Omega_1 \cap \rho^{-1}(V)$.
(vii) $\rho(\tilde{\sigma}_\tau[s_0, t_\tau])$ is embedded.
(viii) $\tilde{\sigma}_\tau$ avoids the inverse image of the ρ image of S^1.
(ix) $\tilde{\sigma}_\tau(t)$ avoids the zeros of g' for $t < t_\tau$. This can be done since the zeros of g' in the interior are countable. Take any $\tilde{\sigma}_\tau$, then one has uncountably many disjoint variations fixing q, and thus, infinitely many miss the zeros of g'.

Let $\hat{\sigma}_\tau$ be an analytic approximation to $\tilde{\sigma}_\tau$ so that $g(\hat{\sigma}_\tau[0, s_0])$ is an embedding, $\hat{\sigma}_j$ a root curve close to σ_j for $0 \leq s \leq s_0$, $\hat{\sigma}_\tau[0, t_\tau]$ and with (iv) and (vi)–(ix) holding for $\hat{\sigma}_\tau$. By (7.5), $\rho(\hat{\sigma}_\tau(t)) = \rho(\hat{\sigma}_j(t))$ for all t. Thus for $t \geq s_0$, $\hat{\sigma}_j(t) \in \rho^{-1}(V) \subset W$. Since $\hat{\sigma}_\tau[0, t_\tau]$ is embedded, it follows that $\hat{\sigma}_j[0, t_\tau]$ is also embedded.

As $s \to t_\tau$, let $q_1 \in \overline{W}$ be any limit point. Then

$$\rho(q_1) = (q_1 - q)^r \tilde{\rho}(q_1) + c = \rho(q) + c = c.$$

Thus $q = q_1$. If we define $\hat{\sigma}_j(t_\tau) = q$, then $\hat{\sigma}_j$ is continuous on $[0, t_\tau]$. Now $\hat{\sigma}_j$ and $\hat{\sigma}_\tau$ may be viewed as the boundary of an open disk $\mathcal{D}$ with boundary component 0 and q, and $\hat{\sigma}_j(0, t_\tau)$ and $\hat{\sigma}_\tau(0, t_\tau)$. From (7.5) it follows that if we look at the images of the intervals $(0, t_\tau)$ by $\hat{\sigma}_j$ and $\hat{\sigma}_\tau$ and the values of $\hat{X}$ on these images we obtain

$$\hat{X} \mid \hat{\sigma}_j(0, t_\tau) = \hat{X} \mid \sigma_\tau(0, t_\tau)$$

and since this is true for all admissible variations $\hat{\sigma}_s$ of $\hat{\sigma}$ we see that for t, $0 < t < t_\tau$ the normal derivatives of $\hat{X}$ also agree. By identifying $\hat{\sigma}_j(t)$ and $\hat{\sigma}_\tau(t)$ for $0 \leq t \leq t_\tau$, $\hat{X}(\mathcal{D})$ is a smooth closed minimal submanifold of $\mathbb{R}^3$ with boundary points $\hat{X}(0)$ and $\hat{X}(q)$. Thus, by the convex hull property of minimal surfaces, $\hat{X}(\mathcal{D})$ lies within the convex hull of $\hat{X}(0)$ and $\hat{X}(q)$, i.e. on the straight line joining $\hat{X}(0)$ and $\hat{X}(q)$, a clear impossibility, and Lemma 7.10 is proved. □

7.2 Tomi's Proof of the Gulliver–Osserman–Royden Theorem

F. Tomi has generously allowed us to include his proof of the theorem of Gulliver–Osserman–Royden. His proof does not use the fact that X is a minimum. For the sake of generality, Tomi works with a domain which is a Riemann surface M with k boundary components. Thus, he considers minimal surfaces of genus greater than one. For those unfamiliar with these concepts, assume $k = 1$, $M = B$.

Let M be a compact Riemann surface with boundary ∂M and $X : M \to \mathbb{R}^3$ a minimal surface spanning a collection $\Gamma = \Gamma_1 \cup \cdots \cup \Gamma_k$ of pairwise disjoint Jordan curves $\Gamma_1, \ldots, \Gamma_k$ of class $C^{1,\mu}$, i.e. X is harmonic and conformal (up to branch points) in $\text{int}(M)$, continuous on M, and $X : \partial M \to \Gamma$ is bijective. It follows then that $X \in C^{1,\mu}(M)$ and that X is immersed on M up to finitely many branch points in the interior and on the boundary of M (Nitsche [1]).

We follow the papers of Gulliver as to the basic definitions and the terminology.

Definition 7.1

(i) Two different points $p, q \in \text{int}(M)$ are called equivalent if they are both regular for X and there are disjoint open neighbourhoods V of p and W of q and a diffeomorphism $h : V \to W$ such that $X(p) = X(q)$ and $X|W = (X|V) \circ h$.

(ii) The surface X is called ramified if it admits a pair of equivalent points. A point $p \in M$ is called a ramified point if each neighbourhood of p contains a pair of equivalent points.

(iii) A set $S \subset \partial M$ is defined as the set of all points $p \in \partial M$ such that there is a point $q \in \text{int}(M)$ and a sequence of equivalent pairs (p_n, q_n), $n \in \mathbb{N}$, in $\text{int}(M)$ with the property that $p_n \to p$, $q_n \to q$ $(n \to \infty)$.

Since the tangent planes of the surface X coincide in equivalent points and the tangent planes extend smoothly into branch points, the tangent planes also are identical in limit points of equivalent points, whether the limit points are regular or not.

We would like to show that under suitable assumptions ramification does not occur. We thus exclude false branch points in particular since such points are clearly ramified points. We shall prove the following theorem which is a special case of Gulliver's Theorem 8.9 in (Gulliver [1]). Our proof takes serious advantage of the differentiability of the surface up to the boundary.

Theorem 7.3 *Let $\Gamma = \Gamma_1 \cup \cdots \cup \Gamma_k$ be a collection of pairwise disjoint Jordan curves in $\mathbb{R}^3$ of class $C^{1,\mu}$, $0 < \mu < 1$, and such that each Γ_i contains at least one extreme point of Γ, i.e. a point in the boundary of the convex hull of Γ relative to the affine subspace of $\mathbb{R}^3$ of minimal dimension containing Γ. Then any minimal surface $X : M \to \mathbb{R}^3$ such that $X : \partial M \to \Gamma$ is bijective is not ramified.*

Corollary 7.1 *If Γ is a single Jordan curve of class $C^{1,\mu}$ then any minimal surface spanning Γ is not ramified.*

The theorem (and hence the corollary) is an easy consequence of the following

Proposition 7.2 *If X is ramified then there is a component C_i of ∂M such that each point of C_i is either in S or it is a branch point of X.*

Let us quickly deduce the theorem from the proposition: let Γ_i be the component of Γ such that $\Gamma_i = X(C_i)$. By assumption, Γ_i contains an extreme point p_i of Γ, $p_i = X(z_i)$ with $z_i \in C_i$ and hence $z_i \in S$ or z_i is a branch point. Let then P be a supporting plane in p_i for Γ and hence for $X(M)$. If $z_i \in S$ we may apply the strong maximum principle for harmonic functions and in case that z_i is a branch point Hopf's boundary point lemma to conclude that $X(M)$ is contained in P, in particular $\Gamma \subset P$. Identifying P with $\mathbb{R}^2$ we may repeat the argument in one dimension lower, implying that Γ is contained in a line, a contradiction.

Proof of Proposition 7.2 Let us recall the local behaviour of a minimal surface around a branch point. To begin with let $p \in \text{int}(M)$ be such a point. We introduce a system of Cartesian coordinates (x_1, x_2, x_3) in $\mathbb{R}^3$ such that $X(p)$ is the origin and the plane $x_3 = 0$ is the tangent plane Π of X in $X(p)$. Then, with respect to a local conformal coordinate system $w = u + iv$ around p one has the representation.

$$\begin{aligned} X_u &= a \operatorname{Re} w^m + b \operatorname{Im} w^m + o(w^m), \\ X_v &= -a \operatorname{Im} w^m + b \operatorname{Re} w^m + o(w^m) \end{aligned} \tag{7.11}$$

where $a, b \in \mathbb{R}^3$ with $|a| = |b| > 0$, $\langle a, b\rangle = 0$, $a_3 = b_3 = 0$ and $m \in \mathbb{N}$ is the order of the branch point. Considering a, b as vectors in $\mathbb{R}^2$ we obtain from (7.11) by integration

$$\hat{X} := (X_1, X_2) = \frac{1}{m+1}(a \operatorname{Re} w^{m+1} + b \operatorname{Im} w^{m+1}) + o(w^{m+1}). \tag{7.12}$$

From (7.11) and (7.12) we conclude that there is a disk $D_r = \{(x_1, x_2) \in \Pi \,|\, x_1^2 + x_2^2 < r^2\}$ and a neighbourhood U of $p \in M$ such that $X(U\backslash\{p\})$ is an $(m+1)$-sheeted graph over D_r, i.e. $\hat{X}$ is regular on $U\backslash\{p\}$, $\hat{X}(U\backslash\{p\}) = D_r\backslash\{0\}$ and the degree of $\hat{X}|U\backslash\{p\}$ is $m+1$ on $D_r\backslash\{0\}$. Corresponding representations hold in case p is a boundary branch point, but with the following modification: the local conformal (u, v)-coordinate system is located in the upper half-plane $v \geq 0$ so that $v = 0$ corresponds to ∂M, the order m of the branch point is even, and the orthogonal projection of $X(U \cap \partial M)$ on Π divides the disk D_r into two regions such that the degree of $\hat{X}$ is $\frac{1}{2}m + 1$ on one of them and $\frac{1}{2}m$ on the other one. □

Lemma 7.12 *If X is ramified then the set S is not empty.*

Proof Let $p, q \in \operatorname{int}(M)$ be a pair of equivalent points and let us choose a path $\alpha : [0, 1] \to M$ such that $\alpha([0, 1)) \subset \operatorname{int}(M)$, $\alpha(0) = p$, $\alpha(1) \in \partial M$ and such that α avoids the finite set $X^{-1}(X(B))$, where B denotes the set of branch points of X. Let us consider a maximal corresponding path $\beta : [0, \tau) \to \operatorname{int}(M)$, $0 < \tau \leq 1$, such that $\beta(0) = q$ and $(\alpha(t), \beta(t))$ is a pair of equivalent points for $t \in [0, \tau)$, in particular $\alpha(t) \neq \beta(t)$.

Case 1. $\tau < 1$. If there is a sequence $t_n \to \tau$ $(n \to \infty)$ and a point $w \in \partial M$ with $\beta(t_n) \to w$, then clearly $w \in S$ since $\alpha(t_n) \to \alpha(\tau) \in \operatorname{int}(M)$. In the other case that no such sequence (t_n) exists, the curve β remains in a compact subset of $\operatorname{int}(M)$ and there is a sequence $s_n \to \tau$ and a point $w \in \operatorname{int}(M)$ such that $\beta(s_n) \to w (n \to \infty)$. Since $X(w) = \lim X(\beta(s_n)) = \lim X(\alpha(s_n)) = X(\alpha(\tau))$, w must be a regular point of X since otherwise $\alpha(\tau) \in X^{-1}(X(B))$. Since X is injective in a neighbourhood of $\alpha(\tau)$ it follows that $\alpha(\tau) \neq w$ and since both points are regular for X there are disjoint open neighbourhoods U of $\alpha(\tau)$ and V of w such that $X(U)$ as well as $X(V)$ are graphs over some common domain in the common tangent plane of $X(\alpha(\tau))$ and $X(w)$. Since $\alpha(s_n) \in U$ and $\beta(s_n) \in V$ for sufficiently large n the two graphs coincide on some open set and hence are identical, in particular $X(U) = X(V)$. It follows that $\beta(t) = (X|V)^{-1}(X(\alpha(t)))$ for $t \in [s_n, \tau)$ and that $\beta(t)$ can be extended beyond τ such that $\alpha(t)$ and $\beta(t)$ are equivalent. So β was not maximal and we have shown that S is not empty provided $\tau < 1$.

Case 2. $\tau = 1$. Again, we consider first the case that for some sequence $t_n \to 1$ one has $\beta(t_n) \to w$ $(n \to \infty)$ for some $w \in \partial M$. Because of $X(\alpha(1)) = \lim X(\alpha(t_n)) = \lim X(\beta(t_n)) = X(w)$ one has $\alpha(1) = w$ since $X|\partial M$ is injective. But then $\alpha(1) = w$ is a branch point since X could not be injective on any neighbourhood of $\alpha(1) = w$, contradicting the choice of α. Therefore $\beta(t_n) \to w \in \operatorname{int}(M)$ for some w and some sequence $t_n \to 1$, which shows that $\alpha(1) \in S$. □

Lemma 7.13 *S is open in ∂M.*

Proof Let $p \in S$. According to the definition of S there are $q \in \text{int}(M)$ and a sequence of equivalent pairs (p_n, q_n) such that $p_n \to p$, $q_n \to q$ $(n \to \infty)$. The points p, q may be regular or not but their tangent planes coincide and there are disjoint neighbourhoods U of p and V of q and a disk D_r of radius r and centre $x_0 = X(p) = X(q)$ in the common tangent plane such that the representations (7.11) and (7.12) hold in U and V and $X(U)$ as well as $X(V)$ are multigraphs over $D_r \backslash \{x_0\}$, respectively, as described above. Since $p_n \in U$ and $q_n \in V$ for large n there is an open subset Ω of $D_r \backslash \{x_0\}$ such that some sheet of $X(U)$ coincides with some sheet of $X(V)$ over Ω. By analyticity (or ellipticity) we conclude that $X(U)$ is completely contained in $X(V)$. Let now $z \in (U \backslash \{p\}) \cap \partial M$ and (z_n) a sequence in $(U \backslash \{p\}) \cap \text{int}(M)$ with $z_n \to z$ and let us choose neighbourhoods U_n of z_n, $U_n \subset (U \backslash \{p\}) \cap \text{int}(M)$ such that $X(U_n)$ is a single-valued graph over some open set $\Omega_n \subset D_r \subset \backslash \{x_0\}$. By what we have shown, $X(U_n)$ is contained in some sheet of $X(V)$ and hence coincides with some graph $X(V_n)$, where V_n is a suitable component of $(\hat{X}|V)^{-1}(\Omega_n)$. Defining $w_n := (X|V_n)^{-1}(X(z_n))$, clearly z_n and w_n are equivalent and, passing to a subsequence, we may assume that $w_n \to w \in \bar{V} \subset \text{int}(M)$, showing that $z \in S$. □

Once the following lemma is proved, Proposition 7.2 and hence Theorem 7.3 are established.

Lemma 7.14 *A boundary point p of S in ∂M is a branch point and an isolated point of $\partial M \backslash S$.*

Proof The set S being open as shown in Lemma 7.13, S is a denumerable union of open intervals in ∂M. A boundary point of S therefore either is an endpoint of one of the intervals forming S or it is a limit point of such endpoints. Below we shall show that each such endpoint is a branch point of X, implying that there are only finitely many endpoints and hence only finitely many components of S. Then, of course, each boundary point of S is an endpoint of some subinterval of S. Accordingly, we now consider a boundary point p of some component of S. Then there is a sequence (p_n) in S and a sequence of equivalent pairs (z_n, w_n) in $\text{int}(M)$ such that $p_n \to p$ $(n \to \infty)$ and $d(z_n, p_n) < \frac{1}{n}$ for some metric d on M. Eventually passing to a subsequence, we may also assume that $w_n \to w \in M$ $(n \to \infty)$. In case $w \in \text{int}(M)$, p would belong to S. Therefore $w \in \partial M$ and since $X(w) = \lim X(w_n) = \lim X(z_n) = X(p)$ it follows that $w = p$ and hence p is a branch point since X is not one-to-one on any neighbourhood of p. We now employ the representations (7.11) and (7.12) with the corresponding normalizations. We may furthermore assume that all points $(u, 0) \in U$ with $u > 0$ belong to S. We must show that this also holds for $(u, 0) \in U$ with $u < 0$ provided that U is appropriately chosen. Recall that $\hat{X}(U) = D_r$, a disk in the tangent plane of radius r and centre $0 = X(0)$. From (7.11) and (7.12), using $\langle a\ b \rangle = 0$ and $|a|^2 = |b|^2$ one computes

$$\begin{aligned}\frac{\partial}{\partial\rho}|\hat{X}(\rho e^{i\theta})|^2 &= 2\langle\hat{X}, \hat{X}_\rho\rangle \\ &= \frac{2}{m+1}\,\frac{1}{\rho}\left(|a|^2\,\rho^{2(m+1)}\right) + o\left(\rho^{2(m+1)}\right) > 0 \qquad (7.13)\end{aligned}$$

for $z = \rho e^{i\theta} \in U$ provided that the radius r and hence U are sufficiently small. Equation (7.13) shows that the level sets $|\hat{X}| = s,\ 0 < s < r$, are curves of class C^1, in fact radial graphs with respect to the (ρ, θ)-coordinates and, since $|\hat{X}| = r$ on $\partial U \cap \operatorname{int}(M)$, each level set $|\hat{X}| = s$ connects two points $(u_s^+,\ 0)$ and $(u_s^-,\ 0)$ on $U \cap \partial M$, where $u_s^+ > 0$ and $u_s^- < 0,\ 0 < s < r$. Hence we may parametrize each level set as a curve

$$\begin{aligned}&\alpha_s : [0,1] \to U, \quad \alpha_s(0) = (u_s^+,\ 0), \quad \alpha_s(1) = (u_s^-, 0), \\ &\qquad \alpha_s([0,1]) = \{z \in U \mid |\hat{X}(z)| = s\}, \quad 0 < s < r.\end{aligned} \qquad (7.14)$$

We now come to the central argument of our proof. Let $z = (u, 0) \in U \cap \partial M$ with $u > 0$ be given, $|\hat{X}(z)| = s \in (0, r)$, i.e. $u = u_s^+$. We showed in the proof of Lemma 7.14 that when $z \in S$ the points z_n in a corresponding sequence of equivalent pairs (z_n, w_n) with $z_n \to z$ may be chosen arbitrarily, if only sufficiently close to Z. We use this freedom of choice by choosing z_n on the level curve $|\hat{X}| = s$, i.e. $z_n = \alpha_s(t_n)$ with $t_n > 0,\ t_n \to 0\ (n \to \infty)$. Since the corresponding equivalent points w_n also are in U and satisfy $X(w_n) = X(z_n)$, it follows from (7.14) that $w_n = \alpha_s(\tau_n)$ for some $\tau_n > 0$. It cannot be that $\tau_n < t_n$ for infinitely many n, because then $w_n \to \alpha_s(0) = z$ for a subsequence, implying that z is a branch point, contradicting the fact that 0 is the only branch point of X in U. We may therefore assume that $0 < t_n < \tau_n$ for all n. For some fixed $N \in \mathbb{N}$ let us now repeat the construction of a maximal curve β_s pointwise equivalent to $\alpha_s|[t_N, 1]$, i.e. $\beta_s : [t_N, \tau) \to \operatorname{int}(M)$, $\beta_s(t_N) = w_n = \alpha_s(\tau_N)$ and $(\alpha_s(t), \beta_s(t))$, are equivalent pairs for all $t \in [t_N, \tau)$. Since $X(\beta_s(t)) = X(\alpha_s(t))$, β_s parametrizes a portion of α_s, as follows from (7.14), and hence

$$\beta_s(t) = \alpha_s(\varphi(t))$$

for some continuous φ with $\varphi(t_N) = \tau_N > t_N$. We claim that

$$\varphi(t) > t, \qquad t \in [t_N, \tau). \qquad (7.15)$$

If there were a first value σ with $\varphi(\sigma) = \sigma$ one could choose an arbitrary sequence $\sigma_n \to \sigma\ (n \to \infty)$, $\sigma_n < \sigma$, producing a sequence of equivalent pairs $(\alpha_s(\varphi(\sigma_n)), \alpha_s(\sigma_n))$ such that $\alpha_s(\sigma_n) \to \alpha_s(\sigma)$, $\alpha_s(\varphi(\sigma_n)) \to \alpha_s(\sigma)\ (n \to \infty)$. Thus $\alpha_s(\sigma)$ is a branch point in $U \backslash \{0\}$, a contradiction. Finally, we would like to show that β_s reaches the endpoint of α_s, i.e.

$$\lim_{t \to \tau} \varphi(t) = 1. \qquad (7.16)$$

If not, there is a sequence $t_n \to \tau\ (n \to \infty)$ such that $\varphi(t_n) \to T\ (n \to \infty)$ with $T < 1$. It follows from (7.15) that $\tau \le T$. In case $\tau = T$ we obtain the same contradiction as above, namely that $\alpha_s(\tau)$ ought to be a branch point. If, on the other hand, we had $\tau < T$ then the pair $(\alpha_s(\tau), \alpha_s(T))$ of different regular points in $\text{int}(M)$ is the limit of the sequence of equivalent pairs $(\alpha_s(t_n), \alpha_s(\varphi(t_n)) = \beta_s(t_n))$, $n \to \infty$, and the curve β_s could be extended beyond τ by the same argument already used in the proof of Lemma 7.12. Since τ is maximal, (7.16) and consequently

$$\lim_{t \to \tau} \alpha_s(\varphi(t)) = \alpha_s(1) \tag{7.17}$$

are proved. Repeating an argument from above we can exclude that $\tau = 1$ because then $\alpha_s(1)$ is a branch point. Therefore we have $\tau < 1$, $\alpha_s(\tau) \in \text{int}(M)$ and by (7.16) there is a sequence of equivalent pairs $(\alpha_s(t_n), \beta_s(t_n) = \alpha_s(\varphi(t_n)))$ such that $\alpha_s(t_n) \to \alpha_s(\tau)$, $\beta_s(t_n) \to \alpha_s(1)$. This proves that $\alpha_s(1) \in S$. Since $\alpha_s(1) = (u_s^-, 0)$ with $u_s^- < 0$ and $s \in (0, r)$ may be arbitrarily chosen we have shown that some interval $(-\varepsilon, 0)$ with $\varepsilon > 0$ belongs to S. $\square$

Chapter 8
Boundary Branch Points

In this chapter we first show that Dirichlet's integral possesses intrinsic second and third derivatives at a minimal surface $\hat{X}$ on the tangent space $T_X M$ of $M := H^2(\partial B, \mathbb{R}^n)$ of $X = \hat{X}|_{\partial B}$ on the space $J(\hat{X})$ of forced Jacobi fields for $\hat{X}$. In particular it will be seen that $J(\hat{X})$ is a subspace of the kernel of the Hessian $D^2 E(X)$ of Dirichlet's integral $E(X)$ defined in (8.1) below, and an interesting formula (see (8.16)) for the second variation of Dirichlet's integral is derived.

Secondly we prove that, for a sufficiently smooth contour Γ in $\mathbb{R}^3$, not only the *order*, but also the *index* of a boundary branch point of a minimal surface $X \in \mathcal{C}(\Gamma)$ can be estimated in terms of the *total curvature* of Γ if curvature and torsion of Γ are nowhere zero.

Then we prove Wienholtz's theorem, which states a condition under which a minimizer for Plateau's problem cannot possess a boundary branch point. In particular we show: If n is the order and m the index of a boundary branch point of $\hat{X}$ such that $2m - 2 < 3n$ (equivalently $2m + 2 \leq 3(n + 1)$) then $\hat{X}$ cannot be a minimizer of Dirichlet's integral of area. The key idea of the proof will be to again compute the third derivative of Dirichlet's integral, D, in an intrinsic way on $J(\hat{X})$, thereby showing that the formula for $E^{(3)}(0) = \frac{d^3}{dt^3} D(\hat{Z}(t))\big|_{t=0}$ derived in Chap. 2 is valid in the presence of boundary branch points as well.

Finally, we show that in the presence of a non-exceptional boundary branch point, a minimal surface spanning a sufficiently smooth contour with non-zero curvature and torsion cannot be a minimum for either energy or area.

Towards these goals, we first show that if the boundary contour $\Gamma \subset \mathbb{R}^n$ is of class $C^{r+7}, r \geq 3$, the space $\mathcal{H}^{5/2}_{\Gamma}(\overline{B}, \mathbb{R}^n)$ of harmonic surfaces from $\overline{B}$ into $\mathbb{R}^n$, mapping $S^1 = \partial B$ to Γ, is a C^r manifold, in fact, a C^r-submanifold of the space $\mathcal{H}^{5/2}(\overline{B}, \mathbb{R}^n)$ of harmonic mappings from $\overline{B}$ into $\mathbb{R}^n$. Instead of the dimension $n = 3$ we do this for arbitrary dimension n. Here it is essential that we operate in the context of a manifold since the third derivative of any real-valued C^3-smooth function is seen to be well defined as a trilinear form on the kernel of the Hessian of this function at any critical point. We shall use the symbol D for the total derivative or the Fréchet derivative. Therefore we need another notation for Dirichlet's integral;

A. Tromba, *A Theory of Branched Minimal Surfaces*,
Springer Monographs in Mathematics,
DOI 10.1007/978-3-642-25620-2_8,

instead of D we employ the symbol E and consider E as a function of boundary values $X : S^1 \to \mathbb{R}^n$ (instead of their harmonic extension $\hat{X}$), i.e.

$$E(X) := \frac{1}{2} \int_B (\hat{X}_u \cdot \hat{X}_u + \hat{X}_v \cdot \hat{X}_v)\, du\, dv \quad \text{for } X \in H^{1/2}(S^1, \mathbb{R}^n). \tag{8.1}$$

It is a well-known fact that $\mathbb{R}^n$ carries a C^{r+6}-Riemannian metric g with respect to which Γ is totally geodesic, i.e. any g-geodesic $\sigma : (-1, 1) \to \mathbb{R}^n$ with $\sigma(0) \in \Gamma$ and $\sigma'(0) \in T_{\sigma(0)}\Gamma$ remains on Γ. Let $(p, v) \mapsto \exp_p v$ denote the exponential map of g; it is of class C^{r+4}. Via harmonic extension we identify the space

$$M := H^2(S^1, \Gamma)$$

of H^2-maps from S^1 to Γ with the space $\mathcal{H}^{5/2}_\Gamma(\overline{B}, \mathbb{R}^n)$. In order to show that M is a submanifold of $H^2(S^1, \mathbb{R}^n)$ we need to identify the tangent space $T_X M$ for $X \in H^2(S^1, \Gamma)$.

Definition 8.1 We define the tangent space $T_X M$ of M at $X \in H^2(S^1, \Gamma)$ as

$$T_X M := \{Y \in H^2(S^1, \mathbb{R}^n) : Y(e^{i\theta}) \in T_{X(e^{i\theta})}\Gamma, \quad \theta \in \mathbb{R}\}.$$

Clearly $T_X M$ is a Hilbert subspace of $H^2(S^1, \mathbb{R}^n)$. Our goal is to show that the map

$$\Phi(Y)(s) := \exp_{X(s)} Y(s), \quad s = e^{i\theta},$$

is a local C^r-diffeomorphism about the zero $0 \in H^2(S^1, \mathbb{R}^n)$ mapping a neighbourhood of zero in $T_X M$ onto a neighbourhood of X in M. Towards this goal we have:

Theorem 8.1 *If $\varphi \in C^{r+3}(\mathbb{R}^n, \mathbb{R}^n)$, then $\Phi : H^2(S^1, \mathbb{R}^n) \to H^2(S^1, \mathbb{R}^n)$ defined by $\Phi(Y) := \varphi \circ Y$ is of class C^r. Furthermore,*

$$D^m \Phi_Y(\lambda_1, \ldots, \lambda_m)(s) = D^m \varphi_{Y(s)}(\lambda_1(0), \ldots, \lambda_m(s)) \quad \textit{for } 0 \le m \le r.$$

The proof of this theorem will be a consequence of the following

Lemma 8.1 *Let $\mathcal{L}^m(\mathbb{R}^n, \mathbb{R}^n)$ be the space of m-linear maps from $\mathbb{R}^n$ into $\mathbb{R}^n$, and suppose that $f \in C^3(\mathbb{R}^n, \mathcal{L}^m(\mathbb{R}^n, \mathbb{R}^n))$. Then the map $F : H^2(S^1, \mathbb{R}^n) \to \mathcal{L}^m(H^2(S^1, \mathbb{R}^n), H^2(S^1, \mathbb{R}^n))$ defined by*

$$Y \mapsto F(Y)(\lambda_1, \ldots, \lambda_m)(s) := f(Y(s))(\lambda_1(s), \ldots, \lambda_m(s))$$

is continuous. Moreover, if $f \in C^4$ then $F \in C^1$, and the derivative of $Y \mapsto F(Y)$ is

$$\lambda \mapsto df(Y(s))(\lambda(s), \lambda_1(s), \ldots, \lambda_m(s)).$$

Proof Recall that $H^2(S^1, \mathbb{R}^n)$ is continuously and compactly embedded into $C^1(S^1, \mathbb{R}^n)$. Assume for simplicity that

$$\|\lambda_j\|_{H^2} \le 1, \qquad \|Y\|_{H^2} < 2, \qquad \|\tilde{Y}\|_{H^2} < 2,$$

and consider the difference

$$[F(Y) - F(\tilde{Y})](\lambda_1, \dots, \lambda_m)(s) = [f(Y(s)) - f(\tilde{Y}(s))](\lambda_1(s), \dots, \lambda_m(s)).$$

Then

$$\begin{aligned}
&\frac{d}{ds}[F(Y) - F(\tilde{Y})](\lambda_1, \dots, \lambda_m)(s) \\
&\quad = df(Y(s))(Y'(s))(\lambda_1(s), \dots, \lambda_m(s)) - df(\tilde{Y}(s))(\tilde{Y}'(s))(\lambda_1(s), \dots, \lambda_m(s)) \\
&\qquad + \sum_{j=1}^{m}[f(Y(s)) - f(\tilde{Y}(s))](\lambda_1(s), \dots, \lambda_{j-1}(s), \lambda_j'(s), \lambda_{j+1}(s), \dots, \lambda_m(s)) \\
&\quad = df(Y(s))(Y'(s) - \tilde{Y}'(s))(\lambda_1(s), \dots, \lambda_m(s)) \\
&\qquad + [df(Y(s)) - df(\tilde{Y}(s))](\tilde{Y}'(s))(\lambda_1(s), \dots, \lambda_m(s)) \\
&\qquad + \sum_{j=1}^{m}[f(Y(s)) - f(\tilde{Y}(s))](\lambda_1(s), \dots, \lambda_j'(s), \dots, \lambda_m(s)).
\end{aligned}$$

Since f is Lipschitz continuous, we have

$$\begin{aligned}
\sup_s |f(Y(s)) - f(\tilde{Y}(s))| &\le \text{const} \sup_s |Y(s) - \tilde{Y}(s)| \\
&\le \text{const}\, \|Y - \tilde{Y}\|_{H^1},
\end{aligned}$$

and therefore

$$\begin{aligned}
&\left| \sum_{j=1}^{m}[f(Y(s)) - f(\tilde{Y}(s))](\lambda_1(s), \dots, \lambda_j'(s), \dots, \lambda_m(s)) \right| \\
&\quad \le \text{const} \sum_{j=1}^{m} \|Y - \tilde{Y}\|_{H^1} |\lambda_j'(s)|,
\end{aligned}$$

from which it follows that

$$\left\| \sum_{j=1}^{m}[f(Y) - f(\tilde{Y})](\lambda_1, \dots, \lambda_j', \dots, \lambda_m) \right\|_{L^2} \le \text{const}\, \|Y - \tilde{Y}\|_{H^2}.$$

Furthermore, the Lipschitz continuity of df implies

$$\begin{aligned}
&\|df(Y)(Y' - \tilde{Y}')(\lambda_1, \dots, \lambda_m)\|_{L^2} \le \text{const}\, \|Y - \tilde{Y}\|_{H^2}, \\
&\|[df(Y) - df(\tilde{Y})](\tilde{Y}')(\lambda_1, \dots, \lambda_m)\|_{L^2} \le \text{const}\, \|Y - \tilde{Y}\|_{H^2}.
\end{aligned}$$

Summarizing these estimates we obtain

$$\left\| \frac{d}{ds}[F(Y) - \tilde{F}(Y)](\lambda_1, \dots, \lambda_m) \right\|_{L^2} \le \text{const}\, \|Y - \tilde{Y}\|_{H^2}.$$

In the same manner we infer

$$\left\| \frac{d^2}{ds^2}[F(Y) - \tilde{F}(Y)](\lambda_1, \dots, \lambda_m) \right\|_{L^2} \le \text{const}\, \|Y - \tilde{Y}\|_{H^2},$$

since f, df, and $d^2 f$ are Lipschitz continuous, using

$$
\begin{aligned}
&\frac{d^2}{ds^2}[F(Y) - F(\tilde{Y})](\lambda_1, \dots, \lambda_m)(s) \\
&\quad = d^2 f(Y(s))(Y'(s))(Y'(s) - \tilde{Y}'(s))(\lambda_1(s), \dots, \lambda_m(s)) \\
&\qquad + df(Y(s))(Y''(s) - \tilde{Y}''(s))(\lambda_1(s), \dots, \lambda_m(s)) \\
&\qquad + \sum_{j=1}^{m} df(Y(s))(Y'(s) - \tilde{Y}'(s))(\lambda_1(s), \dots, \lambda_j'(s), \dots, \lambda_m(s)) \\
&\qquad + [d^2 f(Y(s))(Y'(s)) - d^2 f(\tilde{Y}(s))(\tilde{Y}'(s))](\tilde{Y}'(s))(\lambda_1(s), \dots, \lambda_m(s)) \\
&\qquad + [df(Y(s)) - df(\tilde{Y}(s))](\tilde{Y}''(s))(\lambda_1(s), \dots, \lambda_m(s)) \\
&\qquad + \sum_{j=1}^{m} [df(Y(s)) - df(\tilde{Y}(s))](Y'(s))(\lambda_1(s), \dots, \lambda_j'(s), \dots, \lambda_m(s)) \\
&\qquad + \sum_{j=1}^{m} [f(Y(s)) - f(\tilde{Y}(s))](\lambda_1(s), \dots, \lambda_j''(s), \dots, \lambda_m(s)) \\
&\qquad + \sum_{j,k=1, j<k}^{m} [f(Y(s)) - f(\tilde{Y}(s))](\lambda_1(s), \dots, \lambda_j'(s), \dots, \lambda_k'(s), \dots, \lambda_m(s)) \\
&\qquad + \sum_{j=1}^{m} [df(Y(s))(Y'(s)) - df(\tilde{Y})(\tilde{Y}'(s))](\lambda_1(s), \dots, \lambda_j'(s), \dots, \lambda_m(s)).
\end{aligned}
$$

The estimates above prove that F maps $H^2(S^1, \mathbb{R}^n)$ continuously into the space

$$\mathcal{L}^m(H^2(S^1, \mathbb{R}^n), H^2(S^1, \mathbb{R}^m)).$$

If $f \in C^4$ then $df \in C^3$ and $d^2 f \in C^2$, and Taylor's theorem yields

$$f(u + h) - f(u) - df(u)h = r(u, h)(h, h)$$

where

$$r(u, h)(h, h) := \int_0^1 (1 - t)[d^2 f(u + th) - d^2 f(u)](h, h)\, dt.$$

Since f is in C^4 we obtain

$$\|r(u, h)(h, h)\|_{H^2} \le \text{const}\, \|h\|_{H^2}^2 \quad \text{for } \|h\|_{H^2} \le 1.$$

This shows that the mapping F is differentiable, and its derivative $DF(Y)$ at $Y \in H^2(S^1, \mathbb{R}^n)$ is given by

$$(DF(Y)h)(s) = df(Y(s))h(s).$$

Since $df \in C^3$, the first part of the lemma yields $DF \in C^0$. □

Proof of Theorem 8.1 Applying Lemma 8.1 to $f = d^m\varphi$ successively to $m = 0, 1, \dots, r - 1$, we infer that $D\Phi, D^2\Phi, \dots, D^r\Phi$ exist and are continuous. □

Theorem 8.2 *$M = H^2(S^1, \Gamma)$ is a C^r-submanifold of $H^2(S^1, \mathbb{R}^n)$.*

Proof Since $H^2(S^1, \mathbb{R}^n) \subset C^1(S, \mathbb{R}^n)$, the set M is closed in $H^2(S^1, \mathbb{R}^n)$. Consider the map $Y \mapsto \Phi(Y)$ defined by

$$\Phi(Y)(s) := \exp_{X(s)} Y(s) \quad \text{for } X \in H^2(S^1, \Gamma),$$

which is of class C^r by virtue of Theorem 8.1.

Since $\Phi(0)$ is the identity map, the inverse function theorem implies that Φ is a local C^r-diffeomorphism about 0. Moreover, as the Riemannian metric g is totally geodesic with respect to Γ, we see that Φ maps $T_X M$ into M. Since Φ is also locally invertible, it provides a coordinate chart for M as a submanifold of $H^2(S^1, \mathbb{R}^n)$. □

Before we can apply the preceding results to Plateau's problem we need an abstract functional analytic reasoning which shows that a C^3-function $E : M \to \mathbb{R}$ on a C^r-smooth submanifold M of a Hilbert space $\mathcal{H}$, $r \geq 3$, possesses intrinsic first, second, and third order derivatives for any critical point x of E (i.e. $DE(x) = 0$). To prove this we need a few prerequisites.

By $E \in C^3(M)$ we mean that E extends to a C^3-map on a neighbourhood of every point $x \in M$. Equivalently we can use coordinate charts as follows. From the definition of a submanifold it follows that about each point $x \in M$ there is a C^r-diffeomorphism $\rho : \mathcal{V} \to \mathcal{V}'$ from a neighbourhood $\mathcal{V}$ of x in $\mathcal{H}$ onto a neighbourhood $\mathcal{V}'$ of 0 in $\mathcal{H}$ with $\rho(x) = 0$ such that $\rho(\mathcal{V} \cap M)$ is an open subset of a fixed subspace $\mathcal{H}_0$ of $\mathcal{H}$. Then "$E \in C^3(M)$" means that $E \circ \psi$ is of class C^3 for any such chart $(\rho, \mathcal{V})$ where ψ is the inverse of ρ. For $x \in M$ with the image $0 = \rho(x)$ we define the tangent space $T_x M$ of M at x by

$$T_x M := D\psi(0)[\mathcal{H}_0] \subset \mathcal{H},$$

i.e. as the image of $\mathcal{H}_0$ under the mapping provided by the derivative $D\psi(0)$. This definition of $T_x M$ does not depend on the choice of the chart $(\rho, \mathcal{V})$.

As each $h \in T_x M$ can be written as $h = D\psi(0)\tilde{h}$ with $\tilde{h} \in \mathcal{H}_0$, we define

$$DE(x)h := D(E \circ \psi)(0)\tilde{h},$$

which again can be shown to be independent of the choice of the chart.

A point $x \in M$ is a critical point of $E : M \to \mathbb{R}$ if $DE(x) = 0$. At a critical point x of E there is a well-defined bilinear form

$$D^2 E(x) : T_x M \times T_x M \to \mathbb{R}$$

defined by

$$D^2 E(x)(h, k) := D^2(E \circ \psi)(0)(\tilde{h}, \tilde{k})$$
$$\text{for} \quad h = D\psi(0)\tilde{h}, \quad k = D\psi(0)\tilde{k}; \quad \tilde{h}, \tilde{k} \in \mathcal{H}_0.$$

This is the Hessian (bilinear form), which again does not depend on the choice of the chart $(\rho, \mathcal{V})$, as we will shortly show. Surprisingly, there is also a third intrinsic derivative $D^3 E(x)$, but this is intrinsically defined only on the kernel K_x of $D^2 E(x)$, i.e. on

$$K_x := \{h \in T_x M : D^2 E(x)(h, k) = 0 \text{ for all } k \in T_x M\}.$$

Let us state this formally as

Theorem 8.3 *At a critical point x of $E \in C^3(M)$ there is an intrinsically defined*[1] *second derivative $D^2E(x) : T_xM \times T_xM \to \mathbb{R}$, and a third derivative $D^3E(x) : K_x \times K_x \times K_x \to \mathbb{R}$ defined as a trilinear map on the kernel K_x of $D^2E(x)$.*

To prove this we have to show that, with respect to any transition map $\varphi : U \to U$ on $U \subset M$ fixing the critical point $x \in U$ of E, the second and third derivatives of $E \circ \varphi$ depend only on the first derivative of φ and are independent of $D^2\varphi(x)$ and $D^3\varphi(x)$. Since we may choose the critical point x as the origin 0, the theorem is a consequence of the following

Lemma 8.2 *Let U be an open subset of a Hilbert space and suppose that $0 \in U$ is a critical point of $E \in C^3(U)$. Assume also that K is the kernel of the Hessian of E at 0 and $\varphi : U \to U$ is a C^3-diffeomorphism of U onto itself with $\varphi(0) = 0$. Then*

$$D^2(E \circ \varphi)(0)(k_1, k_2) = D^2E(0)(D\varphi(0)k_1, D\varphi(0)k_2),$$

and furthermore, if $D\varphi(0)k_j \in K$, $j = 1, 2, 3$, then

$$D^3(E \circ \varphi)(0)(k_1, k_2, k_3) = D^3E(0)(D\varphi(0)k_1, D\varphi(0)k_2, D\varphi(0)k_3).$$

Proof Repeatedly using the chain rule we see that

(i) $D(E \circ \varphi)(x)(h) = DE(\varphi(x))D\varphi(x)h.$

(ii)
$$\begin{aligned} D^2(E \circ \varphi)(x)(h, k) &= D^2E(\varphi(x))(D\varphi(x)h, D\varphi(x)k) \\ &\quad + DE(\varphi(x))D^2\varphi(x)(h, k). \end{aligned}$$

(iii)
$$\begin{aligned} D^3(E \circ \varphi)(x)(h, k, \ell) &= D^3E(\varphi(x))(D\varphi(x)h, D\varphi(x)k, D\varphi(x)\ell) \\ &\quad + D^2E(\varphi(x))(D^2\varphi(x)(h, \ell), D\varphi(x)k) \\ &\quad + D^2E(\varphi(x))(D\varphi(x)h, D^2\varphi(x)(k, \ell)) \\ &\quad + D^2E(\varphi(x))(D^2\varphi(x)(h, k), D\varphi(x)\ell) \\ &\quad + DE(\varphi(x))D^3\varphi(x)(h, k, \ell). \end{aligned}$$

Set $k_1 := h, k_2 := k, k_3 := \ell$ and note that $DE(0) = 0$. Then the first assertion follows from (ii) and $\varphi(0) = 0$. The second claim is a consequence of (iii) noting that $\varphi(0) = 0$, $DE(0) = 0$, and by assumption $D\varphi(0)k_j \in K$, $1 \le j \le 3$. □

Now we shall apply the preceding result to Dirichlet's integral $E : H^2(S^1, \mathbb{R}^n) \to \mathbb{R}$ defined by (8.1). Recall the assumption $\Gamma \in C^{r+7}, r \ge 3$. By Theorem 8.2 it

[1] An intrinsic derivative $D^*f(x)$ of a map $f : M \to \mathbb{R}$ on a subspace σ of the tangent space T_xM is an r-linear form $\sigma^+ \to \mathbb{R}$ of $\sigma^r = \sigma \times \cdots \times \sigma$ which is defined independently of the choice of any coordinate chart.

follows that $M := H^2(S^1, \Gamma)$ is a $C^{\boldsymbol{r}}$-submanifold of $H^2(S^1, \mathbb{R}^n)$, and since $E : H^2(S^1, \mathbb{R}^n) \to \mathbb{R}$ is of class C^∞, it follows immediately that the restriction $E|M$ is of class $C^{\boldsymbol{r}}$. Let us simply write E instead of $E|M$, i.e. we view E as a function of class $C^r(M)$.

We wish now to calculate the intrinsic third derivative in the direction of certain specific elements of the kernel of $D^2E(X) : T_X M \times T_X M \to \mathbb{R}$, namely the forced Jacobi fields, in the case that $X \in H^2(S^1, \Gamma)$ is a minimal surface. By the results of Stefan Hildebrandt we know that $\hat{X} \in C^{\boldsymbol{r}+6,\alpha}(\overline{B}, \mathbb{R}^n)$ and therefore also $X \in C^{\boldsymbol{r}+6,\alpha}(S^1, \mathbb{R}^n)$ for all $\alpha \in (0, 1)$.

Besides assuming that $\Gamma \in C^{\boldsymbol{r}+7}$ we make another standing assumption on Γ, namely that the total curvature $\int_\Gamma \kappa\, ds$ of Γ satisfies

$$\int_\Gamma \kappa\, ds \leq \frac{1}{3}\pi \boldsymbol{r}, \tag{8.2}$$

which implies $\boldsymbol{r} \geq 6$. Then the generalized Gauss–Bonnet formula (Wienholtz [3]) implies

$$2\pi \sum_{w_j \in B} \nu(w_j) + \pi \sum_{\zeta_k \in \partial B} \nu(\zeta_k) + 2\pi \leq \frac{1}{3}\pi \boldsymbol{r}$$

where $\nu(w_j)$ are the orders of the interior branch points w_j of a (branched) minimal surface $\hat{X} \in \mathcal{C}(\Gamma)$, and $\nu(\zeta_k)$ are the orders of its boundary branch points, $k = 1, \ldots, q$. Suppose that $q \geq 1$. Then

$$\nu(\zeta_k) \leq \boldsymbol{r}/3 - 2. \tag{8.3}$$

Recall the definition of a *forced Jacobi field* of a minimal surface $\hat{X} : \overline{B} \to \mathbb{R}^3$ which we now generalize to a minimal surface $\hat{X} : \overline{B} \to \mathbb{R}^n$ with $n \geq 3$ which has the interior branch points $w_1, \ldots, w_p$ and the boundary branch points $\zeta_1, \ldots, \zeta_q$. The *generator* τ of a forced Jacobi field $\hat{Y}$ for $\hat{X}$ is a meromorphic function on $\overline{B}$ with poles possibly at $w = 0$ and at the branch points of $\hat{X}$ whose orders are at most $\nu(w_j)$ at $w_j \neq 0$, $\nu(0) + 1$ at $w = 0$, $\nu(\zeta_j)$ at ζ_j, and which is real on ∂B. Then the *forced Jacobi field* $\hat{Y}$ of $\hat{X}$ with the generator τ is a mapping $\hat{Y} : \overline{B} \to \mathbb{R}^n$ of the form

$$\hat{Y} = 2\beta \operatorname{Re}(iw\hat{X}_w \tau) \quad \text{with } \beta \in \mathbb{R},$$

and

$$Y = \beta X_\theta \tau|_{S^1} : S^1 \to \mathbb{R}^n$$

are its boundary values. From the regularity of $\hat{X}$ and (8.3) we infer as in Sect. 7.1 that certainly $Y \in H^2(S^1, \mathbb{R}^n)$, $\hat{Y}_w \in C^0(\overline{B}, \mathbb{R}^n)$, and clearly $Y \in T_X M$. The space of forced Jacobi fields of $\hat{X}$ is denoted by $J(\hat{X})$.

We shall show that the forced Jacobi fields are in the kernel of the Hessian of $E : M \to \mathbb{R}$, and we will compute the second and third derivative of E in these directions.

Computation of D^2E and D^3E Let $\Omega(p) : \mathbb{R}^n \to T_p\Gamma$ be the C^{r+6}-smooth orthogonal projection of $\mathbb{R}^n$ onto the tangent space $T_p\Gamma$ for $p \in \Gamma$. We extend $\Omega(p)$ to a C^{r+6}-smooth mapping $p \mapsto \Omega(p)$ from $\mathbb{R}^n$ into $\mathcal{L}(\mathbb{R}^n, \mathbb{R}^n)$. We then can write the first derivative of E at $X \in M = H^2(S^1, \Gamma)$ as

$$DE(X) = \int_{S^1} \langle \Omega(X)\hat{X}_r, h\rangle d\theta, \quad \hat{X}_r = \text{radial derivative of } \hat{X}. \tag{8.4}$$

A slight generalization of Theorem 8.1 yields that $\{X \to \Omega(X)\} \in C^r(M, H^2(S^1, \mathcal{L}(\mathbb{R}^n, \mathbb{R}^n)))$, $M = H^2(S^1, \Gamma)$, if we take Theorem 8.2 into account. Clearly, *X is a critical point of E if and only if*

$$\Omega(X)\hat{X}_r = 0. \tag{8.5}$$

$\hat{X}$ will be a solution to Plateau's problem if X is also a monotonic map from S^1 onto Γ.

The derivative of $\Omega(X)\hat{X}_r$ is given by

$$h \mapsto \Omega(X)\hat{h}_r + D\Omega(X)h[\hat{X}_r], \tag{8.6}$$

and so the Hessian of E is

$$D^2E(X)(h,k) = \int_{S^1} \langle \Omega(X)\hat{h}_r + D\Omega(X)h[\hat{X}_r], k\rangle d\theta. \tag{8.7}$$

It follows that *the kernel of* (8.6) *is just the kernel of the Hessian $D^2E(X)$ of E at X.*

Claim *The forced Jacobi fields of X lie in the kernel of $D^2E(X)$.* To see this we first note that

$$|X_\theta|^2\Omega(X)m = \langle m, X_\theta\rangle X_\theta \quad \text{for } m \in \mathbb{R}^n. \tag{8.8}$$

Differentiating this in the direction of a tangent vector $h \in T_XM$, $M = H^2(S^1, \Gamma)$, we obtain

$$2\langle X_\theta, h_\theta\rangle \Omega(X)[m] + |X_\theta|^2 D\Omega(X)(h)[m] = \langle m, h_\theta\rangle X_\theta + \langle m, X_\theta\rangle h_\theta. \tag{8.9}$$

Thus the kernel of (8.6) is the kernel of

$$h \mapsto |X_\theta|^{-2}\{\langle \hat{X}_r, h_\theta\rangle X_\theta + \langle \hat{X}_r, X_\theta\rangle h_\theta - 2\langle X_\theta, h_\theta\rangle \Omega(X)\hat{X}_r\} + \Omega(X)\hat{h}_r.$$

From (8.5) we infer

$$\langle \hat{X}_r, X_\theta\rangle = 0 \quad \text{and} \quad \Omega(X)\hat{X}_r = 0,$$

and (8.8) yields

$$\Omega(X)\hat{h}_r = |X_\theta|^{-2}\langle \hat{h}_r, X_\theta\rangle X_\theta.$$

Thus h is in the kernel of (8.6) if and only if

$$|X_\theta|^{-2}\{\langle \hat{X}_r, h_\theta\rangle X_\theta + \langle X_\theta, \hat{h}_r\rangle X_\theta\} = 0$$

that is, if and only if

$$\langle \hat{X}_r, h_\theta \rangle + \langle X_\theta, \hat{h}_r \rangle = 0, \tag{8.10}$$

since the zeros of $X_\theta(\theta)$ are isolated because of the asymptotic expansion of $\hat{X}_w$ at branch points $w_0 \in \overline{B}$.

On $S^1 = \partial B$ we have

$$iw\hat{X}_w = \frac{1}{2}(X_\theta + i\hat{X}_r), \qquad iw\hat{h}_w = \frac{1}{2}(h_\theta + i\hat{h}_r),$$

implying that

$$\langle \hat{X}_r, h_\theta \rangle + \langle X_\theta, \hat{h}_r \rangle = -4\,\mathrm{Im}\{w^2 \langle \hat{X}_w, \hat{h}_w \rangle\}. \tag{8.11}$$

If $\hat{h}$ is a forced Jacobi field we have

$$h = \beta X_\theta \tau|_{S^1} \quad \text{and} \quad \hat{h} = 2\,\mathrm{Re}(\beta i w \hat{X}_w \tau)$$

with $\beta \in \mathbb{R}$ and τ the generator of $\hat{h}$. Since $w\hat{X}_w\tau$ is holomorphic on $\overline{B}$, it follows

$$\hat{h}_w = \beta[iw\hat{X}_w\tau]_w.$$

Hence, if $w \in \overline{B}$ is not a branch point of $\hat{X}$, we obtain

$$\hat{h}_w(w) = \beta[i\hat{X}_w(w)\tau + iw\hat{X}_{ww}(w)\tau(w) + iw\hat{X}_w(w)\tau_w(w)].$$

On the other hand, a minimal surface $\hat{X}$ satisfies

$$\langle \hat{X}_w, \hat{X}_w \rangle = 0$$

and therefore also

$$\langle \hat{X}_w(w), \hat{h}_w(w) \rangle = 0$$

if $w \in \overline{B}$ is not a branch point of $\hat{X}$, and by continuity of $\hat{h}_w$ on $\overline{B}$ it follows

$$\langle \hat{X}_w, \hat{h}_w \rangle = 0 \quad \text{if } \hat{h} \in J(\hat{X}). \tag{8.12}$$

From (8.10), (8.11) and (8.12) we infer that for a forced Jacobi field $\hat{h}$ its boundary values h lie in the kernel of (8.6) and therefore in the kernel K_X of the Hessian $D^2E(X)$. This proves the claim, and we have established

Proposition 8.1 *If $\hat{X}$ is a minimal surface with $X \in M = H^2(S^1, \Gamma)$ then the boundary values h of any $\hat{h} \in J(\hat{X})$ lie in the kernel K_X of the Hessian $D^2E(X)$ of E at X, that is, $h \in T_X M$ and*

$$D^2E(X)(h,k) = 0 \quad \text{for all } k \in T_X M.$$

Remark 8.1 We would like to point out that $D^2E(X)$ has been defined for *branched* minimal surfaces without making normal variations of $\hat{X}$.

Before we compute $D^3E(X)$ we give a geometric interpretation of

$$D^2E(X)(h,h) = \delta^2 E(X,h),$$

i.e. of the second variation of E at X in direction of $h \in T_X M$. An integration by parts yields

$$\begin{aligned}\int_{\overline{B}} \nabla\hat{h} \cdot \nabla\hat{h}\, du\, dv &= \int_{S^1} \langle \hat{h}_r, h\rangle d\theta - \int_B \langle \Delta\hat{h}, \hat{h}\rangle du\, dv \\ &= \int_{S^1} \langle \hat{h}_r, h\rangle d\theta \end{aligned} \tag{8.13}$$

since $\Delta\hat{h} = 0$. Away from branch points on S^1 we set

$$h = aX_\theta \quad \text{and} \quad b = \langle \hat{h}_r, X_\theta\rangle.$$

By (8.8) we have

$$\Omega(X)\hat{h}_r = |X_\theta|^{-2}\langle \hat{h}_r, X_\theta\rangle X_\theta,$$

and so

$$\langle h, \Omega(X)\hat{h}_r\rangle = \langle aX_\theta, bX_\theta\rangle |X_\theta|^{-2} = ab = \langle \hat{h}_r, aX_\theta\rangle = \langle \hat{h}_r, h\rangle$$

and by continuity it follows

$$\langle \hat{h}_r, h\rangle = \langle h, \Omega(X)\hat{h}_r\rangle \quad \text{on } S^1.$$

On account of (8.7) and (8.13) it follows that

$$D^2E(X)(h,h) = \int_B |\nabla\hat{h}|^2 du\, dv + \int_{S^1} \langle h, D\Omega(X)h[\hat{X}_r]\rangle d\theta. \tag{8.14}$$

In order to simplify the boundary term we return to (8.9) where we insert $m = \hat{X}_r$. Since $\langle \hat{X}_r, X_\theta\rangle = 0$ we have $\Omega(X)\hat{X}_r = 0$ on S^1, and so two terms in (8.9) vanish. We are left with

$$D\Omega(X)h[\hat{X}_r] = |X_\theta|^{-2}\langle \hat{X}_r, h_\theta\rangle X_\theta.$$

Since $h = aX_\theta$ (away from branch points), we have

$$h_\theta = aX_{\theta\theta} + a_\theta X_\theta$$

whence

$$\langle \hat{X}_r, h_\theta\rangle = a\langle \hat{X}_r, X_{\theta\theta}\rangle.$$

This implies

$$\begin{aligned}\langle h, D\Omega(X)h[\hat{X}_r]\rangle &= |X_\theta|^{-2}\langle aX_\theta, a\langle \hat{X}_r, X_{\theta\theta}\rangle X_\theta\rangle = a^2\langle \hat{X}_r, X_{\theta\theta}\rangle \\ &= |h|^2|X_\theta|^{-2}\langle \hat{X}_r, X_{\theta\theta}\rangle = |h|^2 k_g\end{aligned}$$

where

$$k_g := |X_\theta|^{-2}\langle \hat{X}_r, X_{\theta\theta}\rangle \tag{8.15}$$

is the *signed geodesic curvature of* Γ *in the minimal surface* $\hat{X}$, *i.e. the interior product of the curvature vector of* Γ *with the unit vector* $|\hat{X}_r|^{-1}\hat{X}_r$, since $|X_\theta| = |\hat{X}_r|$ on S^1.

Thus we infer from (8.14) the following result which was independently obtained by R. Böhme and A. Tromba:

Proposition 8.2 *If* $\hat{X}$ *is a minimal surface with* $X \in M = H^2(S^1, \Gamma)$ *then, for any* $h \in T_X M$, *we obtain*

$$D^2E(X)(h,h) = \int_B |\nabla \hat{h}|^2 du\, dv + \int_{S^1} k_g |h|^2 d\theta, \tag{8.16}$$

where k_g *is the signed geodesic curvature* (8.15) *of the boundary contour* Γ *in the minimal surface* $\hat{X}$.

Now we proceed to compute the intrinsic third derivative $D^3E(X)$. Let us return to formula (8.9) which will be differentiated in the direction of a vector $k \in T_X M$. This yields

$$\begin{aligned} &2\langle h_\theta, k_\theta\rangle \Omega(X)m + 2\langle X_\theta, h_\theta\rangle D\Omega(X)[k]m \\ &\quad + 2\langle X_\theta, k_\theta\rangle D\Omega(X)[h]m + |X_\theta|^2 D^2\Omega(X)(h,k)m \\ &= \langle m, h_\theta\rangle k_\theta + \langle m, k_\theta\rangle h_\theta. \end{aligned}$$

Choosing $m := \hat{X}_r$ we see that

$$\begin{aligned} &2\langle X_\theta, h_\theta\rangle D\Omega(X)(k)[\hat{X}_r] + 2\langle X_\theta, k_\theta\rangle D\Omega(X)(h)[\hat{X}_r] \\ &\quad + |X_\theta|^2 D^2\Omega(X)(h,k)[\hat{X}_r] = \langle \hat{X}_r, h_\theta\rangle k_\theta + \langle \hat{X}_r, k_\theta\rangle h_\theta. \end{aligned}$$

By (8.7) we may write for h, k in the kernel of $D^2E(X)$ (and therefore in the kernel of (8.6))

$$D\Omega(X)(h)[\hat{X}_r] = -\Omega(X)\hat{h}_r, \qquad D\Omega(X)(k)[\hat{X}_r] = -\Omega(X)\hat{k}_r, \tag{8.17}$$

then obtaining

$$\begin{aligned} &-2\langle X_\theta, h_\theta\rangle \Omega(X)\hat{k}_r - 2\langle X_\theta, k_\theta\rangle \Omega(X)\hat{h}_r \\ &\quad + |X_\theta|^2 D^2\Omega(X)(h,k)[\hat{X}_r] = \langle \hat{X}_r, h_\theta\rangle k_\theta + \langle \hat{X}_r, k_\theta\rangle h_\theta. \end{aligned} \tag{8.18}$$

Setting in (8.9) $m = \hat{k}_r$ we get

$$2\langle X_\theta, h_\theta\rangle \Omega(X)\hat{k}_r + |X_\theta|^2 D\Omega(X)[h]\hat{k}_r = \langle \hat{k}_r, h_\theta\rangle X_\theta + \langle \hat{k}_r, X_\theta\rangle h_\theta. \tag{8.19}$$

Commuting h and k it follows also

$$2\langle X_\theta, k_\theta\rangle \Omega(X)\hat{h}_r + |X_\theta|^2 D\Omega(X)[k]\hat{h}_r = \langle \hat{h}_r, k_\theta\rangle X_\theta + \langle \hat{h}_r, X_\theta\rangle k_\theta. \tag{8.20}$$

Adding (8.19) and (8.20) to (8.18) we see that

$$\begin{aligned} &|X_\theta|^2 D^2\Omega(X)(h,k)\hat{X}_r + |X_\theta|^2 D\Omega(X)[h]\hat{k}_r + |X_\theta|^2 D\Omega(X)[k]\hat{h}_r \\ &\quad = \langle \hat{X}_r, h_\theta\rangle k_\theta + \langle \hat{X}_r, k_\theta\rangle h_\theta + \langle \hat{h}_r, k_\theta\rangle X_\theta + \langle \hat{h}_r, X_\theta\rangle k_\theta \\ &\qquad + \langle \hat{k}_r, h_\theta\rangle X_\theta + \langle \hat{k}_r, X_\theta\rangle h_\theta. \end{aligned} \tag{8.21}$$

By (8.10) we have

$$\langle X_\theta, \hat{h}_r\rangle = -\langle \hat{X}_r, h_\theta\rangle \quad \text{and} \quad \langle X_\theta, \hat{k}_r\rangle = -\langle \hat{X}_r, k_\theta\rangle.$$

Therefore (8.21) reduces to

$$\begin{aligned}&|X_\theta|^2\{D^2\Omega(X)(h,k)\hat{X}_r + D\Omega(X)[h]\hat{k}_r + D\Omega(X)[k]\hat{h}_r\}\\ &= \{\langle \hat{h}_r, k_\theta\rangle + \langle \hat{k}_r, h_\theta\rangle\}X_\theta.\end{aligned} \tag{8.22}$$

Suppose now that h, k, ℓ lie in the space $J(\hat{X})$ of forced Jacobi fields. By (8.7) we have

$$D^2E(X)(h,\ell) = \int_{S^1}\langle D\Omega(X)h[\hat{X}_r] + \Omega(X)\hat{h}_r, \ell\rangle d\theta. \tag{8.22'}$$

Differentiating this in the direction of k it follows

$$D^3E(X)(h,\ell,k) = \int_{S^1}\langle D^2\Omega(X)(h,k)[\hat{X}_r] + D\Omega[h]\hat{k}_r + D\Omega(X)[k]\hat{h}_r, \ell\rangle d\theta, \tag{8.23}$$

which by (8.22) yields

$$D^3E(X)(h,\ell,k) = \int_{S^1}\{\langle \hat{h}_r, k_\theta\rangle + \langle \hat{k}_r, h_\theta\rangle\}|X_\theta|^{-2}\langle X_\theta, \ell\rangle d\theta. \tag{8.24}$$

Actually there are two more terms on the right-hand side of (8.24) which come from the derivatives ℓ' and h' of ℓ and h. We have to show that these terms are zero if ℓ and h are forced Jacobi fields. The additional ℓ'-term is

$$\int_{S^1}\langle D\Omega(X)h[\hat{X}_r] + \Omega(X)\hat{h}_r, \ell'\rangle d\theta.$$

It vanishes since

$$D\Omega(X)h[\hat{X}_r] + \Omega(X)\hat{h}_r = 0,$$

as h is a forced Jacobi field.

The second additional term becomes

$$\int_{S^1}\langle h', (\widehat{\lambda X_\theta})_r - (\lambda\hat{X}_r)_\theta\rangle d\theta$$

if we write $\ell = \lambda X_\theta = \text{Re}\{\lambda i w \hat{X}_w\}$ and integrate by parts. But ℓ is holomorphic in B and so the Cauchy–Riemann equations yield

$$-\frac{\partial}{\partial\theta}(\widehat{\lambda X_\theta}) + \frac{\partial}{\partial r}(\widehat{\lambda X_\theta}) = 0.$$

This equation extends to the boundary $S^1 = \partial B$, and so the second additional term vanishes too.

The two expressions (8.23) and (8.24) yield the intrinsic third derivative of E at X. We synonymously write

$$\frac{\partial E}{\partial h}(X) = DE(X)h,$$
$$\frac{\partial^2 E}{\partial h \partial k}(X) = D^2 E(X)(h,k), \tag{8.25}$$
$$\frac{\partial^3 E(X)}{\partial h \partial \ell \partial k} = D^3 E(X)(h,\ell,k).$$

Suppose that $h, k, \ell \in J(\hat{X})$ have the generators τ, ρ, λ; we shall write τ, ρ, λ also for the boundary values $\tau|_{S^1}, \rho|_{S^1}, \lambda|_{S^1}$:

$$\begin{aligned} h(\theta) &= \tau(\theta) X_\theta(\theta), \quad \text{so} \quad & \hat{h}(w) &= 2\,\mathrm{Re}(iw\tau(w)\hat{X}_w(w)), \\ k(\theta) &= \rho(\theta) X_\theta(\theta), & \hat{k}(w) &= 2\,\mathrm{Re}(iw\rho(w)\hat{X}_w(w)), \\ \ell(\theta) &= \lambda(\theta) X_\theta(\theta), & \hat{\ell}(w) &= 2\,\mathrm{Re}(iw\lambda(w)\hat{X}_w(w)). \end{aligned} \tag{8.26}$$

Then (8.24) becomes

$$D^3 E(X)(h,\ell,k) = \int_{S^1} \{\langle \hat{h}_r, k_\theta \rangle + \langle \hat{k}_r, h_\theta \rangle\} \lambda(\theta)\, d\theta. \tag{8.27}$$

On S^1 we have $d\theta = \frac{dw}{iw}$ and

$$2w\hat{h}_w = \hat{h}_r - ih_\theta, \qquad 2w\hat{k}_w = \hat{k}_r - ik_\theta$$

whence

$$\langle \hat{h}_r, k_\theta \rangle + \langle \hat{k}_r, h_\theta \rangle = -4\,\mathrm{Im}(w^2 \hat{h}_w \hat{k}_w).$$

Furthermore,

$$\begin{aligned} \hat{h}_w &= (iw\hat{X}_w\tau)_w = i(w\tau\hat{X}_{ww} + \hat{X}_w\tau + w\hat{X}_w\tau_w), \\ \hat{k}_w &= (iw\hat{X}_w\rho)_w = i(w\rho\hat{X}_{ww} + \hat{X}_w\rho + w\hat{X}_w\rho_w). \end{aligned}$$

Since $\hat{X}_w \cdot \hat{X}_w = 0$ and $\hat{X}_w \cdot \hat{X}_{ww} = 0$ it follows that

$$w^2 \hat{h}_w \hat{k}_w = -w^4 \tau\rho \hat{X}_{ww} \cdot \hat{X}_{ww}$$

and consequently

$$\langle \hat{h}_r, k_\theta \rangle + \langle \hat{k}_r, h_\theta \rangle = 4\,\mathrm{Im}(w^4 \tau\rho \hat{X}_{ww} \cdot \hat{X}_{ww}).$$

This implies

$$\begin{aligned} D^3 E(X)(h,\ell,k) &= 4 \int_{S^1} \mathrm{Im}(w^4 \tau\rho \hat{X}_{ww} \cdot \hat{X}_{ww}) \lambda \, d\theta \\ &= 4\,\mathrm{Im} \int_{S^1} w^4 \tau\rho\lambda \hat{X}_{ww} \cdot \hat{X}_{ww}\, d\theta \\ &= 4\,\mathrm{Im} \int_{S^1} w^4 \tau\rho\lambda \hat{X}_{ww} \cdot \hat{X}_{ww} \frac{dw}{iw}, \end{aligned}$$

and we arrive at

$$
\begin{aligned}
D^3 E(X)(h, \ell, k) &= -4 \operatorname{Re} \int_{S^1} w^3 \tau \rho \lambda \hat{X}_{ww} \cdot \hat{X}_{ww} \, dw \\
&= 4 \int_{S^1} \operatorname{Im}(w^4 \tau \rho \lambda \hat{X}_{ww} \cdot \hat{X}_{ww}) \, d\theta. \qquad (8.28)
\end{aligned}
$$

It follows from (8.23) that the right-hand side of (8.28) is the integral of a continuous function. If we wish to apply the residue theorem to evaluate the integral in (8.28) we have to get a better grip on the integrand. To this end we impose an *additional standing assumption*: $n = 3$, *i.e. we consider boundary contours only in* $\mathbb{R}^3$.

First we wish to understand what the generators τ of forced Jacobi fields for a minimal surface $\hat{X}$ with a boundary branch point $w_0 \in S^1$ are. By means of a rotation we can move w_0 to the point $w = 1$. Thus we make the following further standing assumption:

Assumption *$\hat{X} \in \mathcal{C}(\Gamma)$ is a minimal surface in the unit disk B with the boundary branch point $w = 1$ of order n, and the boundary contour $\Gamma \in C^2$ has a total curvature $\kappa(\Gamma) := \int_\Gamma \kappa(s)\, ds$ satisfying $3\kappa(\Gamma) \le \pi \boldsymbol{r}$. It is also assumed that $\Gamma \in C^{\boldsymbol{r}+7}$, $\boldsymbol{r} \ge 2$, which implies $\hat{X} \in C^{\boldsymbol{r}+6,\beta}(\overline{B}, \mathbb{R}^3)$, $0 < \beta < 1$, and $n \le \boldsymbol{r}/3 - 2$.*

It is easy to verify that

$$
\tau(w) := \beta \left(i \frac{w+1}{w-1} \right)^{\ell}, \quad \beta \in \mathbb{R}, \qquad (8.29)
$$

is a meromorphic function on $\overline{B}$ with a pole of order ℓ at $w = 1$ such that $\tau(w) \in \mathbb{R}$ for $w \in S^1 \setminus \{1\}$. If $\ell \le n$ then $\hat{X}_w(w)\tau(w)$ is holomorphic in B and at least continuous on $\overline{B}$ since we have the asymptotic expansion

$$
\begin{aligned}
&\hat{X}_w(w) = a(w-1)^n + o(|w-1|^n) \quad \text{as } w \to 1, \ w \in \overline{B} \setminus \{1\} \\
&\text{with } a \in \mathbb{C}^3, \ a \neq 0, \ \text{and } a \cdot a = 0. \qquad (8.30)
\end{aligned}
$$

Thus τ generates a forced Jacobi field for $\hat{X}$. Consider the conformal mapping $\varphi : \overline{B} \setminus \{-1\} \to \overline{\mathcal{H}}$, defined by

$$
w \mapsto z = \varphi(w) := -i \frac{w-1}{w+1}, \quad w \in \overline{B} \setminus \{-1\}, \qquad (8.31)
$$

which maps $B = \{w \in \mathbb{C} : |w| < 1\}$ onto the upper half-plane

$$
\mathcal{H} := \{z \in \mathbb{C} : \operatorname{Im} z > 0\}
$$

and takes $S^1 \setminus \{-1\}$ onto the real line $\mathbb{R}$ such that $\varphi(1) = 0$, $\varphi(i) = 1$, $\varphi(-1) = \infty$. The inverse $\psi := \varphi^{-1}$ is given by

$$
z \mapsto w = \psi(z) := \frac{1+iz}{1-iz}.
$$

We write $z = x + iy$ with $x = \operatorname{Re} z$ and $y = \operatorname{Im} z$, while $w = u + iv$, $u = \operatorname{Re} w$, $v = \operatorname{Im} w$. From (8.31) we infer

$$\frac{1}{z} = i\frac{w+1}{w-1}$$

and so

$$\sigma := \tau \circ \psi = \frac{\beta}{z^{\ell}}. \tag{8.32}$$

Transforming the minimal surface $\hat{X}(w)$ to the new parameter z, we obtain

$$\hat{Y}(z) := \hat{X}(\psi(z)) \tag{8.33}$$

which has the branch point $z = 0$ on $\mathbb{R} = \partial\mathcal{H}$ with the asymptotic expansion

$$\hat{Y}_z(z) = bz^n + o(|z|^n) \quad \text{as } z \to 0,\ z \in \overline{\mathcal{H}} \setminus \{0\},$$
$$b \in \mathbb{C}^3 \setminus \{0\},\ b \cdot b = 0.$$

Choosing a suitable coordinate system in $\mathbb{R}^3$ we may assume that $\hat{Y}_z(z)$ can be written in the normal form

$$\hat{Y}_z(z) = \tilde{A}_1 z^n + o(z^n) \tag{8.34}$$

with $\tilde{A}_1 = (a_1 + ib_1)$; $a_1, b_1 \in \mathbb{R}^3$, $|a_1|^2 = |b_1|^2 \neq 0$; $a_1 \cdot b_1 = 0$, $a_1 = (n+1)\alpha e_1$, $e_1 = (1, 0, 0)$, $\alpha > 0$, where a_1, b_1 span the tangent space to $\hat{X}$ at $X(1)$. Let us recall that the order of any boundary branch point is even; thus we can set

$$n = 2\nu \quad \text{with } \nu \in \mathbb{N}. \tag{8.35}$$

Now we wish to write $\hat{Y}_z$ in the more specific form

$$\hat{Y}_z(z) = (A_1 z^n + \cdots + A_{m-n+1} z^m + O(|z|^{m+1}),\ R_m z^m + O(|z|^{m+1})) \tag{8.36}$$

with

$$R_m \neq 0. \tag{8.37}$$

By Taylor's theorem and (8.34) we can achieve (8.36) for any $m \in \mathbb{N}$ with $m > n$ and such that $\hat{Y} \in C^{m+2}(\overline{\mathcal{H}}, \mathbb{R}^3)$.

However, it is not at all a priori obvious that one can also achieve (8.37). This fact is ensured by the following

Proposition 8.3 *Suppose that $\hat{Y} \in C^{3n+6}(\overline{\mathcal{H}}, \mathbb{R}^3)$ and that both the torsion $\boldsymbol{\tau}$ and the curvature $\boldsymbol{\kappa}$ of Γ are non-zero. Then there is an $m \in \mathbb{N}$ with $n + 1 < m + 1 \leq 3(n+1)$ such that*

$$\hat{Y}_z^3(z) = R_m z^m + O(|z|^{m+1}) \quad \textit{for } |z| \ll 1 \textit{ and } R_m \neq 0. \tag{8.38}$$

Proof Otherwise we have

$$\hat{Y}_z^3(z) = O(|z|^{3n+3}). \tag{8.39}$$

Let $\gamma(s) = (\gamma_1(s), \gamma_2(s), \gamma_3(s))$ be the local representation of Γ with respect to its arc-length parameter s such that $\gamma(0) = \hat{Y}(0)$ and $\gamma'(0) = e_1$. By (8.34) and (8.35) we have

$$\hat{Y}_x(x, 0) = (n+1)\alpha e_1 x^n + O(x^{n+1}), \quad n = 2\nu,$$

and so s and x are related by $s = \sigma(x)$ with

$$\sigma'(x) = |Y_x(x)| = [(n+1)\alpha x^n + O(x^{n+1})],$$

whence

$$\sigma(x) = \alpha x^{n+1} + O(x^{n+2}) \quad \text{as } x \to 0. \tag{8.40}$$

Then $Y(x) = \gamma(\sigma(x))$ for $|x| \ll 1$, and therefore the third component Y^3 of Y is given by

$$Y^3(x) = \gamma_3(\sigma(x)) = \gamma_3(\alpha x^{n+1} + O(x^{n+2})) \quad \text{for } x \to 0.$$

Because of (8.39) we have $Y_x^3(x) = O(x^{3n+3})$ as $x \to 0$, which implies

$$Y^3(x) = O(x^{3n+4}) \quad \text{as } x \to 0. \tag{8.41}$$

On the other hand

$$\gamma(s) = \gamma'(0)s + O(s^2) \quad \text{as } s \to 0.$$

Consequently

$$Y^3(x) = \gamma_3'(0)\alpha x^{n+1} + O(x^{n+2}) \quad \text{as } x \to 0.$$

On account of (8.41) and $\alpha > 0$ it follows $\gamma_3'(0) = 0$. Thus we can write

$$\gamma_3(s) = \frac{1}{2}\gamma_3''(0)s^2 + O(s^3) \quad \text{as } s \to 0,$$

which implies

$$Y^3(x) = \frac{1}{2}\gamma_3''(0)\alpha^2 x^{2n+2} + O(x^{2n+3}) \quad \text{as } x \to 0.$$

By (8.41) and $\alpha > 0$ we obtain $\gamma_3''(0) = 0$, and we have

$$\gamma_3(s) = \frac{1}{6}\gamma_3'''(0)s^3 + O(s^4) \quad \text{as } s \to 0.$$

Hence,

$$Y^3(x) = \frac{1}{6}\gamma_3'''(0)\alpha^3 x^{3n+3} + O(x^{3n+4}) \quad \text{as } x \to 0,$$

and then $\gamma_3'''(0) = 0$ on account of (8.41) and $\alpha > 0$. Thus we have found

$$\gamma_3'(0) = 0, \qquad \gamma_3''(0) = 0, \qquad \gamma_3'''(0) = 0,$$

and so the three vectors $\gamma'(0), \gamma''(0), \gamma'''(0)$ are linearly dependent. This will contradict our assumption $\kappa(s) \neq 0$ and $\tau(s) \neq 0$. To see this we introduce the Frenet triple $\boldsymbol{T}(s)$, $\boldsymbol{N}(s)$, $\boldsymbol{B}(s)$ of the curve Γ satisfying $\boldsymbol{T} = \gamma'$, $\boldsymbol{T}' = \gamma''$, $\boldsymbol{T}'' = \gamma'''$, and

$$\begin{aligned}
\boldsymbol{T}' &= \quad\;\; \kappa \boldsymbol{N} \\
\boldsymbol{N}' &= -\kappa \boldsymbol{T} \qquad\quad + \tau \boldsymbol{B} \\
\boldsymbol{B}' &= \qquad\; - \tau \boldsymbol{N}.
\end{aligned}$$

Then $\boldsymbol{T}_3(0) = 0$, $\boldsymbol{T}_3'(0) = 0$, $\boldsymbol{T}_3''(0) = 0$, and from $\boldsymbol{T}' = \kappa \boldsymbol{N}$ and $\kappa \neq 0$ it follows that $\boldsymbol{N}_3(0) = 0$. Since

$$\boldsymbol{N}' = \left(\frac{1}{\kappa}\right)' \boldsymbol{T}' + \frac{1}{\kappa} \boldsymbol{T}''$$

we obtain $\boldsymbol{N}_3'(0) = 0$ whence $\boldsymbol{\tau}(0)\boldsymbol{B}_3(0) = 0$. Because of $\boldsymbol{\tau} \neq 0$ it follows that $\boldsymbol{B}_3(0) = 0$, and so $\boldsymbol{T}(0)$, $\boldsymbol{N}(0)$, $\boldsymbol{B}(0)$ are linearly dependent. This is a contradiction since $(\boldsymbol{T}, \boldsymbol{N}, \boldsymbol{B})$ is an orthonormal frame, hence the assumption (8.39) is impossible. □

Remark 8.2 Note that $n \leq \boldsymbol{r}/3 - 2$ implies $3n + 6 \leq \boldsymbol{r} < \boldsymbol{r} + 7$. Thus the assumption $\hat{Y} \in C^{3n+6}(\overline{\mathcal{H}}, \mathbb{R}^3)$ is certainly satisfied if we assume $3\kappa(\Gamma) \leq \pi \boldsymbol{r}$ and $\Gamma \in C^{\boldsymbol{r}+7}$. Thus we have a lower bound on $\boldsymbol{r}$ and upper bounds on n and m. We call the number m in (8.38) with $n < m < 3n + 3$ the *index* of the boundary branch point $z = 0$ of $\hat{Y}$, or of the boundary branch point $w = 1$ of $\hat{X}$.

Assumption *In what follows we assume that the assumptions and therefore also the conclusions of Proposition* 8.3 *are satisfied.*

Proposition 8.4 *If* $m + 1 \not\equiv 0 \bmod (n + 1)$ (*i.e. if* $z = 0$ *is not an exceptional branch point of* $\hat{Y}$) *then the coefficient* R_m *in* (8.38) *satisfies*

$$\operatorname{Re} R_m = 0, \tag{8.42}$$

i.e. R_m *is purely imaginary, and therefore*

$$R_m^2 < 0 \tag{8.43}$$

since $R_m \neq 0$. *If we write* (8.38) *in the form*

$$Y_z^3(z) = R_m z^m + R_{m+1} z^{m+1} + R_{m+2} z^{m+2} + o(|z|^{m+2}) \quad \text{for } |z| \ll 1 \tag{8.44}$$

and if $2m - 2 < 3n$, *then we in addition obtain that*

$$\operatorname{Re} R_{m+1} = 0 \quad \text{and, if } n > 2, \text{ also} \quad \operatorname{Re} R_{m+2} = 0. \tag{8.45}$$

Finally, independent of any assumption on m, *we have*

$$A_j = \mu_j A_1, \quad j = 1, \ldots, \min\{n + 1, 2m - 2n\}, \text{ with } \mu_j \in \mathbb{R} \tag{8.46}$$

for the coefficients A_j *in the expansion* (8.36).

Remark 8.3 The relations (8.46) are in some sense a strengthening of the equations

$$A_j = \lambda_j A_1, \quad j = 1, \ldots, 2m - 2n, \text{ with } \lambda_j \in \mathbb{C}$$

which hold at an interior branch point $w = 0$ of a minimal surface $\hat{X}$ in normal form.

Proof of Proposition 8.4 (i) From (8.44) we infer

$$Y^3(x) = \operatorname{Re}\left(\frac{R_m}{m+1}x^{m+1} + \frac{R_{m+1}}{m+2}x^{m+2} + \frac{R_{m+2}}{m+3}x^{m+3} + o(x^{m+3})\right) \quad \text{for } x \to 0. \tag{8.47}$$

On the other hand,

$$Y^3(x) = \gamma_3(\alpha x^{n+1} + o(x^{n+1}))$$

and $\gamma(0) = 0$, $\gamma'(0) = e_3$ whence also $\gamma_3(0) = \gamma_3'(0) = 0$. As pointed out before it is then impossible that both $\gamma_3''(0) = 0$ and $\gamma_3'''(0) = 0$ because this would imply that $\boldsymbol{T}(0)$, $\boldsymbol{N}(0)$, $\boldsymbol{B}(0)$ are linearly dependent. Thus we obtain

$$\gamma_3(s) = \frac{1}{k!}\gamma^{(k)}(0)s^k + O(s^{k+1}) \quad \text{as } s \to 0,\ \gamma^{(k)}(0) \neq 0,$$

for $k = 2$ or $k = 3$. Therefore

$$Y^3(x) = \frac{1}{k!}\gamma_3^{(k)}(0)\alpha^k x^{k(n+1)} + o(x^{k(n+1)}) \quad \text{as } x \to 0. \tag{8.48}$$

Comparing (8.47) and (8.48) it follows that $\operatorname{Re} R_m \neq 0$ implies $m + 1 = k(n + 1)$ for $k = 2$ or $k = 3$, which is excluded by assumption. Thus $\operatorname{Re} R_m = 0$, and we have

$$\begin{aligned} Y^3(x) &= \operatorname{Re}\left(\frac{R_{m+1}}{m+2}x^{m+2} + \frac{R_{m+2}}{m+3}x^{m+3} + o(x^{m+3})\right) \\ &= \frac{1}{k!}\gamma^k(0)\alpha^k x^{k(n+1)} + o(x^{k(n+1)}) \quad \text{as } x \to 0. \end{aligned} \tag{8.49}$$

Suppose now that $2m - 2 < 3n$, which is equivalent to

$$2m \le 3n \tag{8.50}$$

since n is even, and so

$$m + 2 < m + 3 \le \frac{3}{2}n + 3 < 3(n + 1).$$

Thus, for $k = 3$, (8.49) can only hold if

$$\operatorname{Re} R_{m+1} = 0 \quad \text{and} \quad \operatorname{Re} R_{m+2} = 0.$$

Furthermore, (8.50) yields also

$$m + 2 < m + 3 \le \frac{3}{2}n + 3 = (2n + 2) + \left(1 - \frac{n}{2}\right)\begin{cases} = 2n + 2 & \text{when } n = 2 \\ < 2n + 2 & \phantom{\text{when }} n > 2. \end{cases}$$

Hence it follows in this case that always $\operatorname{Re} R_{m+1} = 0$ while $\operatorname{Re} R_{m+2} = 0$ holds for $n > 2$.

(ii) From $Y_x(x) = 2\operatorname{Re}\hat{Y}_z(x, 0)$, (8.49) and (8.36) it follows that

$$Y_x(x) = 2\operatorname{Re}(A_1 x^n + \cdots + A_{n+1}x^{2n} + o(x^{2n}), o(x^{2n}))$$

whence

$$Y(x) = 2\,\mathrm{Re}\left(\frac{A_1}{n+1}x^{n+1} + \cdots + \frac{A_{n+1}}{2n+1}x^{2n+1} + o(x^{2n+1}), o(x^{2n+1})\right).$$

Furthermore,

$$\gamma(s) = e_1 s + O(s^2) \quad \text{as } s \to 0$$

and

$$\sigma(x) = b_1 x^{n+1} + \cdots + b_{n+1} x^{2n+1} + o(x^{2n+1}) \quad \text{as } x \to 0$$

with $b_1, \ldots, b_{n+1} \in \mathbb{R}$, $\alpha e_1 = b_1 e_1 = \frac{2}{n+1}\,\mathrm{Re}\,A_1$. Then

$$Y(x) = \gamma(\sigma(x)) = (b_1 x^{n+1} + \cdots + b_{n+1} x^{2n+1}) e_1 + O(x^{2n+2}).$$

Comparing the coefficients we get

$$2\,\mathrm{Re}\,A_j = (n+j) b_j e_1 \quad \text{with } \alpha = b_1 > 0 \text{ for } 1 \le j \le n+1.$$

Then $\mathrm{Re}\,A_j = \frac{(n+j)b_j}{(n+1)\alpha}\,\mathrm{Re}\,A_1$, and so

$$\mathrm{Re}\,A_j = \mu_j\,\mathrm{Re}\,A_1 \quad \text{for } j = 2, \ldots, n+1$$

with

$$\mu_j := \frac{n+j}{n+1}\,\frac{b_j}{\alpha}, \qquad 2 \le j \le n+1.$$

Set $A_j := a_j + i b_j$; $a_j := \mathrm{Re}\,A_j$, $b_j := \mathrm{Im}\,A_j \in \mathbb{R}^n$. We know from Lemma 2.2 of Chap. 2 that $A_j = \lambda_j A_1$ for $j = 1, \ldots, 2m - 2n$ with $\lambda_j \in \mathbb{C}$ hence

$$a_j = (\mathrm{Re}\,\lambda_j) a_1 - (\mathrm{Im}\,\lambda_j) b_1 \quad \text{for } 2 \le j \le 2m - 2n$$

and

$$a_j = \mu_j a_1 \quad \text{for } 2 \le j \le n+1.$$

From $|\hat{Y}_x| = |\hat{Y}_y|$ it follows that $|b_1| = |a_1| = \frac{n+1}{2}\alpha > 0$, and $\hat{Y}_x \cdot \hat{Y}_y = 0$ yields $a_1 \cdot b_1 = 0$; thus we obtain $\mathrm{Im}\,\lambda_j = 0$ for $j = 2, \ldots, n+1$ whence $\lambda_j = \mu_j \in \mathbb{R}$ and $A_j = \mu_j A_1$ for $1 \le j \le \min\{n+1, 2m-2n\}$. □

Let us now return to formula (8.28) for $D^3 E(X)(h, k, \ell)$ in the direction of forced Jacobi fields (with the boundary values) h, k, ℓ; note that (8.28) is symmetric in h, k, ℓ. We already know that (8.28) is the integral of a continuous function; but we need to understand (8.28) at a level where we can apply the residue theorem. To this end we consider the conformal mapping (8.31) defined by

$$w \mapsto z = \varphi(w) := -i\,\frac{w-1}{w+1}, \qquad w \in \overline{B} \setminus \{-1\}, \tag{8.51}$$

which has the derivative

$$\varphi'(w) = \frac{-2i}{(w+1)^2}.$$

Using the inverse

$$z \mapsto w = \psi(z) := \frac{1+iz}{1-iz}$$

we obtain

$$\varphi'(\psi(z)) = \frac{-i}{2}(1-iz)^2, \tag{8.52}$$

or sloppily

$$\frac{dz}{dw} = -\frac{i}{2}(1-iz)^2.$$

From (8.33) we get $\hat{X}(w) = \hat{Y}(\varphi(w))$, whence

$$\hat{X}_{ww} = \hat{Y}_{zz}(\varphi)(\varphi')^2 + \hat{Y}_z(\varphi)\varphi''.$$

From $\hat{Y}_z \cdot \hat{Y}_z = 0$ it follows $\hat{Y}_z \cdot \hat{Y}_{zz} = 0$, and then

$$\hat{X}_{ww} \cdot \hat{X}_{ww} = \hat{Y}_{zz}(\varphi) \cdot \hat{Y}_{zz}(\varphi)(\varphi')^4, \tag{8.53}$$

which we sloppily write

$$\hat{X}_{ww} \cdot \hat{X}_{ww} = \hat{Y}_{zz} \cdot \hat{Y}_{zz} \left(\frac{dz}{dw}\right)^4.$$

Lemma 8.3 *Assuming* $2m-2 < 3n$ *(i.e.* $2m \le 3n$*) we obtain the Taylor expansion*

$$(\hat{Y}_{zz} \cdot \hat{Y}_{zz})(z) = \sum_{j=0}^{s} Q_j z^{2m-2+j} + R(z) \tag{8.54}$$

with $s := (3n-1) - (2m-2) = (3n-2m) + 1 \ge 1$, $R(z) = O(z^{3n})$, *where* $Q_0 := (m-n)^2 R_m^2 < 0$ *and* $\operatorname{Im} Q_j = 0$ *for* $0 \le j \le s$.

Proof From $2m-2 < 3n$ we infer $2m \le 3n$ since n is even. Thus $s \ge 1$ and $2m - 2n + 1 \le n+1$. Consider the Taylor expansion

$$\hat{Y}_z(z) = (A_1 z^n + A_2 z^{n+1} + \cdots, R_m z^m + R_{m+1} z^{m+1} + \cdots)$$

where "$+\cdots$" indicates further z-powers plus a remainder term. As for interior branch points we have

$$A_1 \cdot A_{2m-2n+1} = -R_m^2/2 \tag{8.55}$$

and

$$A_2 \cdot A_{2m-2n+1} + A_1 \cdot A_{2m-2n+2} = -R_m R_{m+1}. \tag{8.56}$$

By (8.43) we have $R_m^2 < 0$ whence $A_1 \cdot A_{2m-2n+1} \in \mathbb{R}$. Since $2 \le 2m - 2n \le n$ it follows $A_2 = \mu_2 A_1$ with $\mu_2 \in \mathbb{R}$ on account of (8.46). Then (8.55) implies $A_2 \cdot A_{2m-2n+1} \in \mathbb{R}$, and furthermore $R_m R_{m+1} \in \mathbb{R}$ in virtue of (8.43) and (8.45). Then (8.56) yields $A_1 \cdot A_{2m-2n+2} \in \mathbb{R}$, and we arrive at

$$\hat{Y}_{zz}(z) \cdot \hat{Y}_{zz}(z) = Q_0 z^{2m-2} + Q_1 z^{2m-1} + \cdots$$

with $Q_0 = (m-n)^2 R_m^2$, and $Q_0 < 0$ as well as $Q_1 \in \mathbb{R}$, since Q_1 is a real linear combination of $A_1 \cdot A_{2m-2n+2}$, $A_2 \cdot A_{2m-2n+1}$, and $R_m R_{m+1}$. Suppose now that $s = 3n - 2m + 1 > 1$. In order to show $\operatorname{Im} Q_j = 0$ for $2 \le j \le s$, we note that by (8.53)

$$\tau\rho\lambda w^4 \hat{X}_{ww} \cdot \hat{X}_{ww} = \tau\rho\lambda \hat{Y}_{zz} \cdot \hat{Y}_{zz} \left(w \frac{dz}{dw} \right)^4,$$

where τ, ρ, λ are generators of forced Jacobi fields with the pole $w = 1$. Furthermore, by (8.51),

$$w \frac{dz}{dw} = \frac{-2iw}{(w+1)^2} = \frac{1+z^2}{2i}. \tag{8.57}$$

Thus

$$\operatorname{Im}(\tau\rho\lambda w^4 \hat{X}_{ww} \cdot \hat{X}_{ww}) = \frac{1}{16} \operatorname{Im}[\tau\rho\lambda(1+z^2)^4 \hat{Y}_{zz} \cdot \hat{Y}_{zz}]. \tag{8.58}$$

By (8.28) the left-hand side of (8.58) is a continuous function on S^1, and thus the right-hand side must be continuous in a neighbourhood of 0 in $\overline{\mathcal{H}}$ for all generators τ, ρ, λ of forced Jacobi fields $\hat{h}, \hat{k}, \hat{l}$ with poles at $w = 1$.

Suppose now that not all Q_j with $2 \le j \le s$ are real, $s = (3n-1) - (2m-2)$, and let J be the smallest of the indices $j \in \{2, \dots, s\}$ with the property that $\operatorname{Im} Q_j \ne 0$. Then we choose λ, ρ, τ such that the sum of their pole orders at $w = 1$ equals $(J+1) + (2m-2) \le 3n$. Transforming λ, ρ, τ from w to z it follows for $z = x \in \mathbb{R} = \partial\mathcal{H}$ that

$$\begin{aligned} &\operatorname{Im}[\tau\rho\lambda(1+z^2)^4 \hat{Y}_{zz} \cdot \hat{Y}_{zz}]\Big|_{z=x\in\mathbb{R}} \\ &= (1+x^2)^4 \beta_1 (\operatorname{Im} Q_J) \frac{1}{x} + \langle \text{terms continuous in } x \rangle, \end{aligned} \tag{8.59}$$

$\beta_1 \in \mathbb{R} \setminus \{0\}$. This is clearly not a continuous function unless $\operatorname{Im} Q_J = 0$, a contradiction, therefore no such J exists. □

Now we want to evaluate the integral in (8.28) by applying the residue theorem to this and we state

Proposition 8.5 *Let τ be given by* (8.29), *and consider the function*

$$f(w) := \tau(w)^4 w^4 \hat{X}_{ww}(w) \cdot \hat{X}_{ww}(x), \quad w \in \overline{B}, \tag{8.60}$$

which has a continuous imaginary part on $S^1 = \partial B$. Then there is a meromorphic function $g(w)$ on $\overline{B}$ with a pole only at $w = 1$ such that

(i) $\operatorname{Im}[f(w) - g(w)] = 0$ *for* $w \in S^1 = \partial B$;
(ii) $f - g$ *is continuous on* $\overline{B}$.

Proof Setting $w = \psi(z) = (1+iz)/(1-iz)$ we obtain

$$f(\psi(z)) = \frac{1}{16} \tau(\psi(z))^3 (1+z^2)^4 \hat{Y}_{zz}(z) \cdot \hat{Y}_{zz}(z).$$

By (8.54) of Lemma 8.3 we see that, in a neighbourhood of $z = 0$ in $\mathcal{H}$, we can write the right-hand side as

$$\sum_{j=0}^{s} \sum_{\ell_j} \tilde{\beta}_j \tilde{Q}_j z^{-l_j} + G(z)$$

with $\tilde{\beta}_j \in \mathbb{R}$, $\tilde{Q}_j \in \mathbb{R}$, $0 < l_j \le (3n - 1) - (2m - 2) = s$, and a continuous term $G(z)$. Set

$$\tilde{g}(z) := \sum_{j=0}^{s} \sum_{l_j=1}^{s} \tilde{\beta}_j \tilde{Q}_j z^{-l_j} \quad \text{for } z \in \overline{\mathcal{H}} \setminus \{0\}$$

and

$$g(w) := \tilde{g}(\varphi(w)) = \sum_{j=0}^{s} \sum_{l_j=1}^{s} \tilde{\beta}_j \tilde{Q}_j \left(i \frac{w+1}{w-1} \right)^{l_j}.$$

Clearly f and g satisfy (i) and (ii). □

Corollary 8.1 *We have*

$$\int_{S^1} [f(w) - g(w)]\, d\theta = -2\pi \operatorname{res}_{w=0} \frac{g(w)}{w}. \tag{8.61}$$

Proof For $w = e^{i\theta} \in S^1$ we have $d\theta = dw/(iw)$, whence

$$\begin{aligned} \int_{S^1} [f(w) - g(w)]\, d\theta &= \int_{S^1} [f(w) - g(w)] \frac{dw}{iw} \\ &= 2\pi \operatorname{res}_{w=0} \left\{ \frac{f(w) - g(w)}{w} \right\} \\ &= -2\pi \operatorname{res}_{w=0} \left\{ \frac{g(w)}{w} \right\} \end{aligned}$$

since $f(w)/w$ is holomorphic at $w = 0$. □

Since $\operatorname{Im} g = 0$ on S^1, we obtain

Corollary 8.2 *We have*

$$\operatorname{Im} \int_{S^1} f(w)\, d\theta = 2\pi \operatorname{Im} \operatorname{res}_{w=0} \left\{ \frac{g(w)}{w} \right\}. \tag{8.62}$$

Furthermore we have

$$\begin{aligned} -4 \operatorname{Re}\{w^3 \tau^3 \hat{X}_{ww} \cdot \hat{X}_{ww}\, dw\} &= (-4) \operatorname{Re}\{i w^4 \tau^3 \hat{X}_{ww} \cdot \hat{X}_{ww}\}\, d\theta \\ &= 4 \operatorname{Im}\{w^4 \tau^3 \hat{X}_{ww} \cdot \hat{X}_{ww}\}\, d\theta = 4 \operatorname{Im} f(w)\, d\theta. \end{aligned}$$

Then (8.28) and Corollary 8.2 imply

$$D^3 E(X)(h, h, h) = -8\pi \operatorname{Im} \operatorname{res}_{w=0} \left\{ \frac{g(w)}{w} \right\}. \tag{8.63}$$

Remark 8.4 We note the following slight, but very useful generalization of the three preceding results. Namely, if $\hat{X}$ has other boundary branch points than $w=1$ we are allowed to change τ by an additive term having poles of first order at these branch points. Then Proposition 8.5 as well as Corollaries 8.1 and 8.2 also hold for the new f defined by (8.60) and the modified τ. This observation is used in order to ensure that the forced Jacobi field $\hat{h}$ generated by τ produces a variation $\hat{Z}(t)$, $|t| \ll 1$, of $\hat{X}$ which is monotonic on $\partial B = S^1$.

Now we turn to the evaluation of $D^3E(X)(h,h,h)$ using formula (8.63). We distinguish three possible cases: There is an $l \in N$ such that

(i) $2m-1=3l$, then l is odd;
(ii) $2m-2=3l$, in this case l is even;
(iii) $2m=3l$, here l is again even.

Since $2m \le 3n$ it follows $l<n$ for (i) and (ii), whereas $l \le n$ in case (iii).

Case (i). Choose τ as

$$\begin{aligned} &\tau := \beta\tau_1 + \epsilon\tau^* \quad \text{and} \quad \beta > 0, \quad \epsilon > 0, \quad \text{and} \\ &\tau_1 = \left(i\frac{w+1}{w-1}\right)^l = \frac{1}{z^l}, \quad w \in \overline{B}\setminus\{1\}, \end{aligned} \tag{8.64}$$

$w=\psi(z)$, $w \in \overline{B}\setminus\{-1\}$, $z \in \overline{\mathcal{H}}\setminus\{0\}$. We will choose τ^* as a meromorphic function that has poles of order 1 at the boundary branch points different from $w=1$ or $z=0$ respectively. Then close to $w=1$ or $z=0$ respectively we have

$$\begin{aligned} \tau^3 w^4 \hat{X}_{ww}\cdot\hat{X}_{ww} &= \frac{1}{16}\tau^3(1+z^2)^4\hat{Y}_{zz}\cdot\hat{Y}_{zz} \\ &\overset{(8.54)}{=} \frac{\beta^3}{16}(m-n)^2 R_m^2 \frac{1}{z} + G(z) + O(\epsilon) \end{aligned}$$

with a continuous $G(z)$.

Choose

$$g(w) = \frac{\beta^3}{16}(m-n)^2 R_m^2 \left(i\frac{w+1}{w-1}\right)$$

and let $\hat{h}(w) = \operatorname{Re}(iw\hat{X}_w(w)\tau(w))$ be the forced Jacobi field generated by τ, $h := \hat{h}|_{S^1}$. Then by Proposition 8.5 and Corollaries 8.1 and 8.2 we obtain

$$\begin{aligned} D^3E(X)(h,h,h) &= -\frac{8\pi}{16}\beta^3(m-n)^2 R_m^2 \operatorname{Im}\left\{\operatorname{res}_{w=1}\frac{i}{w}\left(\frac{w+1}{w-1}\right)\right\} \\ &= \frac{1}{2}\pi\beta^3(m-n)^2R_m^2 + O(\epsilon). \end{aligned} \tag{8.65}$$

Since $R_m^2 < 0$ this yields for $0<\epsilon\ll 1$ that

$$D^3E(X)(h,h,h) < 0.$$

Case (ii). Here we have $3l = 2m - 2 < 3n$ whence $l < n$. Since both l and n are even we obtain $l + 1 < n$ whence $n > 2$. Moreover, $2m - 1 = 2(l + 1) + (l - 1)$. Set

$$\begin{aligned} &\tau := \epsilon\tau_1 + \beta\tau_2 + \epsilon^3\tau^*, \quad \beta > 0, \quad \epsilon > 0, \\ &\tau_1 := \left(i\frac{w+1}{w-1}\right)^{l+1}, \quad \tau_2 := \left(i\frac{w+1}{w-1}\right)^{l-1}, \quad \tau^* \quad \text{as in Case (i).} \end{aligned} \tag{8.66}$$

Note also that both $l + 1$ and $l - 1$ are odd. We then have that

$$\begin{aligned} \tau^3 &= \beta^3\tau_2^3 + 3\beta^2\tau_2^2\tau_1\epsilon + 3\beta\epsilon^2\tau_1^2\tau_2 + O(\epsilon^3) \\ &= \beta^3 z^{-2m+5} + 3\beta^2 z^{-2m+3} + 3\beta\epsilon^2 z^{-2m+1} + O(\epsilon^3) \end{aligned}$$

for z close to zero, but this does not add a contribution to (8.63).

By the same procedure as in Case (i) we find for $\hat{h} = \text{Re}(iw\hat{X}_w\tau)$ that

$$D^3E(X)(h,h,h) = \frac{3}{2}\pi\epsilon^2\beta(m-n)^2R_m^2 + O(\epsilon^3), \tag{8.67}$$

which implies

$$D^3E(X)(h,h,h) < 0 \quad \text{for } 0 < \epsilon \ll 1.$$

Case (iii). Now we have $2m = 3l$, $l =$ even. We have two subcases.

(a) If $l = n$ we write $2m - 1 = 2l + (l - 1)$ and set

$$\begin{aligned} &\tau_1 := \left(i\frac{w+1}{w-1}\right)^{l-1}, \quad \tau_2 := \left(i\frac{w+1}{w-1}\right)^{l}, \\ &\tau := \beta\tau_1 + \epsilon\tau_2 + \epsilon^3\tau^*, \quad \beta > 0, \quad \epsilon^3 > 0. \end{aligned} \tag{8.68}$$

(b) If $l < n$ we write $2m - 1 = 2(l - 1) + (l + 1)$ and set

$$\begin{aligned} &\tau_1 := \left(i\frac{w+1}{w-1}\right)^{l+1}, \quad \tau_2 := \left(i\frac{w+1}{w-1}\right)^{l-1}, \\ &\tau := \epsilon\tau_1 + \beta\tau_2 + \epsilon^3\tau^*. \end{aligned} \tag{8.69}$$

Then our now established procedure yields

$$D^3E(X)(h,h,h) = \begin{cases} \frac{1}{2}3\pi\beta\epsilon^2(m-n)^2R_m^2 + O(\epsilon^3) & \text{in Subcase (a),} \\ \frac{1}{2}3\pi\beta^2\epsilon(m-n)^2R_m^2 + O(\epsilon^2) & \text{in Subcase (b).} \end{cases} \tag{8.70}$$

This again implies $D^3E(X)(h,h,h) < 0$ for $0 < \epsilon \ll 1$ and $\hat{h} = \text{Re}(iw\hat{X}_w\tau)$.

Remark 8.5 The choice of τ^* has to be carried out in such a way that the variation $Z(t)$ of X produced by $\hat{h} = \text{Re}(iw\hat{X}_w\tau)$ furnishes a monotonic mapping of $\partial B = S^1$ onto the boundary contour Γ. The details on how this can be achieved by the formulae (8.64), (8.66), (8.68) and (8.69) can be found in the thesis of D.

Wienholtz [2]. The complete proof is technically quite involved and will be omitted here. We discuss our own approach when considering shortly the fourth, fifth and sixth derivatives at a boundary branch point. We remark here that, at an exceptional branch point, using only the generator τ^*, derivatives of Dirichlet's energy up to order 7 vanish. This observation allows us to ignore τ^* for derivatives up to order 8, for if there is another branch point which is not exceptional one considers the lowest order derivative which is non-negative.

In conclusion we have

Theorem 8.4 (D. Wienholtz) *If $\hat{X}$ is a minimal surface in $\mathcal{C}(\Gamma)$ with $\Gamma \in C^{r+7}, 3\int_\Gamma \kappa\, ds \le \pi \mathbf{r}$, having a boundary branch point of order n and index m satisfying the Wienholtz condition $2m-2<3n$, then X cannot be an $H^2(S^1,\mathbb{R}^3)$-minimizer for Dirichlet's integral $E(X)$ defined by* (8.1), *and thus $\hat{X}$ cannot be an $H^{5/2}(\overline{B},\mathbb{R}^3)$-minimizer of area.*

We now proceed to consider the cases $2m-2\ge 3n$, $m+1\ne k(n+1)$, $k=2,3$. Our plan is to proceed by considering three main additional cases $3n\le 2m-2<4n$, $4n<2m-2\le 5n$, and $5n<2m-2\le 6n$. These will then include all non-exceptional cases $2m+2\le 6n+6$. Within the three main cases there will be sub-cases; e.g. we begin with $2m-2<4(n-1)$. In this regard we need a strengthening of Proposition 8.4:

Proposition 8.6 *If in Proposition* 8.4, $3n\le 2m-2<4n$, *then*

$$\operatorname{Re} R_{m+s}=0, \quad 0\le s\le 2n-m. \tag{8.71}$$

If $2m-2>4n$, *then*

$$\operatorname{Re} R_{m+s}=0, \quad 0\le s\le (3n+1)-m. \tag{8.72}$$

Proof is given below following Lemma 8.6.

We now discuss how to take variations of $\hat{X}$ so that we can calculate both the fourth and fifth derivatives of Dirichlet's energy. For simplicity of exposition we will now be assuming Γ in C^∞ smooth.

In $\mathcal{H}$ we write $z=x+iy$. Again parametrizing Γ by γ we have that

$$\hat{Y}(x)=\gamma\circ\sigma(x)$$

where

$$\sigma(x)=b_1x^{n+1}+\cdots+b_{n+1}x^{2n+1}+o(x^{2n+1}) \quad \text{as } x\to 0. \tag{8.73}$$

We next define a 1-parameter family of maps $\sigma_t, \sigma_0=\sigma$, by

$$\sigma_t(x):=\sigma(x)+t\xi(x)+(t^2/2)\rho(x)$$

where ξ and ρ are C^∞ smooth maps chosen in part, so that $x\mapsto\sigma_t(x)$ remains one to one with $\sigma_t(\pm\infty)=\pm\infty$. Assuming that this is possible, we define

$$\varphi_t(x):=\sigma^{-1}(\sigma_t(x)). \tag{8.74}$$

Clearly φ_t will not be differentiable (in x or t) everywhere. Now define a mapping $\phi(t)$ by

$$\phi(t)(x) := (d\varphi_t/dt)/(d\varphi_t/dx) \tag{8.75}$$

which is defined for those x for which $d\varphi_t/dx \neq 0$. We now have:

Lemma 8.4 *At all points where*

$$\begin{aligned}&\sigma_x + t\xi_x + (t^2/2)\rho_x \neq 0,\\ &\phi(t) = [\xi + t\rho]/[\sigma_x + t\xi_x + (t^2/2)\rho_x].\end{aligned} \tag{8.76}$$

Proof Since $\sigma(\varphi_t(x)) = \sigma_t(x)$ we have

$$\sigma'(\varphi_t(x)) \cdot (d\varphi_t/dt) = d\sigma_t/dt = \xi + t\rho$$

and

$$\sigma'(\varphi_t(x)) \cdot (d\varphi_t/dx) = \sigma_x + \xi_x + (t^2/2)\rho_x.$$

The result now follows. □

We shall see that the computations in Chaps. 2 and 3 of the fourth and fifth derivatives of Dirichlet's energy in the direction of forced Jacobi fields remain valid in the case at hand.

We begin by selecting our variation $\hat{Z}(t)$ of $\hat{X}$. We define $\hat{Z}(t)$ first on the real line in $\overline{\mathcal{H}}$, and then via (8.51) it is defined on S^1, and finally, by harmonic extension to $\overline{B}$.

The mapping $Z(t)$ on $R \subset \overline{\mathcal{H}}$ is defined by

$$Z(t) := \gamma \circ \sigma_t(x) = \gamma \circ \sigma \circ (\sigma^{-1}\sigma_t(x)) = X(\varphi_t(x)) \tag{8.77}$$

and on S^1 via (8.14) and on $\overline{B}$ by harmonic extension. Then $t \mapsto Z(t)$ is C^∞-smooth and $Z(0) = \hat{X}$.

Lemma 8.5

$$\begin{aligned}Z_t := Z'(t) = dZ(t)/dt &= X'(\varphi_t(x))(d\varphi_t/dt)\\ &= X(\varphi_t(x))_x\phi = Z(t)_x\phi.\end{aligned}$$

Proof Chain rule. □

Now φ_t induces a one-parameter family on S^1 via (8.14) and therefore a mapping $\tilde{\phi}$ (introduced as ϕ in Chaps. 2 and 3). We have

Lemma 8.6

$$\begin{aligned}\phi &= \tilde{\phi} i w(dz/dw) = \tilde{\phi}\left(\frac{1+z^2}{2}\right),\\ \phi_t &= \tilde{\phi}_t i w(dz/dw)\end{aligned} \tag{8.78}$$

where ϕ_t denotes the derivative with respect to t.

Proof

$$Z'(t) = \tilde{\phi} Z(t)_\theta$$
$$= Z(t)_x \phi \quad \text{(Lemma 8.2)}$$
$$= \mathrm{Re}(\tilde{\phi} i w \hat{Z}_w) = \mathrm{Re}\left(\tilde{\phi} i w Z_z \frac{dz}{dw}\right)$$
$$= \mathrm{Re}(\hat{Z}(t)_z \phi).$$

□

Proof of Proposition 8.6 We have

$$Y^3(x) = \mathrm{Re}\left(\frac{R_m}{m+1} x^{m+1} + \frac{R_{m+1}}{m+2} x^{m+2} + \frac{R_{m+2}}{m+3} x^{m+3} + o(x^{m+3})\right) \quad \text{for } x \to 0.$$

On the other hand,

$$Y^3(x) = \gamma_3(\alpha x^{n+1} + o(x^{n+1}))$$

and $\gamma(0) = 0$, $\gamma'(0) = e_3$ whence also $\gamma_3(0) = \gamma_3'(0) = 0$. It is then impossible that both $\gamma_3''(0) = 0$ and $\gamma_3'''(0) = 0$. Thus we obtain

$$Y^3(x) = \frac{1}{k!} \gamma_3^{(k)}(0) \alpha^k x^{k(n+1)} + o(x^{k(n+1)}) \quad \text{as } x \to 0.$$

Set $\overline{m} = m + s$. It follows that $\mathrm{Re}\, R_{\overline{m}} \neq 0$ implies $\overline{m} + 1 = k(n+1)$ for $k = 2$ or $k = 3$, which is excluded by assumption. Thus $\mathrm{Re}\, R_{\overline{m}} = 0$. □

Formula (8.32) allows us to use the formulae for the fourth (and later the fifth) derivative of Dirichlet's energy in the direction of forced Jacobi fields; i.e. $\tilde{\phi}$ is a generator of a forced Jacobi field.

We now havc from Chaps. 2 and 3, this time using, for convenience subscripts to denote derivatives in $t, z, \ldots$ etc.

Proposition 8.7 *Assuming $s \geq 4$, $\hat{X} = \hat{Z}(0)$ a minimal surface and $\phi(0) = \tau$, a generator of a forced Jacobi field, the fourth derivative of Dirichlet's energy is given by*

$$\begin{aligned} \frac{d^4 E}{dt^4}(0) &= 12\,\mathrm{Re} \int_{S^1} \hat{Z}_{ttw}(0) \cdot [w \hat{Z}_{tw}(0) \tau + w \hat{X}_w \tilde{\phi}_t(0)]\, dw \\ &\quad + 12\,\mathrm{Re} \int_{S^1} w \hat{Z}_{tw}(0) \cdot \hat{Z}_{tw}(0) \tilde{\phi}_t(0)\, dw. \end{aligned} \tag{8.79}$$

Now restating formula (8.79) on the upper half-plane, we have:

Proposition 8.8 *If $\hat{Z}(t)$, $\hat{X}(t)$, $\hat{Y}$ as above, then*

$$\begin{aligned} \frac{d^4 E}{dt^4}(0) &= 12\,\mathrm{Re} \int \left(w \frac{dz}{dw}\right) \hat{Z}_{ttz}(0) \cdot [\hat{Z}_{tz}(0) \tau + \hat{Y}_z \phi_t(0)]\, d\theta \\ &\quad + 12\,\mathrm{Re} \int \tau^2 \hat{Y}_{zz} \cdot \hat{Y}_{zz} \left(w \frac{dz}{dw}\right) \phi_t(0)\, d\theta. \end{aligned} \tag{8.80}$$

Proof For the second term of (8.80) we have from (8.78)

$$\tilde{\phi}_t(0)\,dw = \phi_t\left(\frac{dw}{dz}\right)\frac{dw}{iw} = \phi_t\left(\frac{dw}{dz}\right)d\theta$$

and

$$\begin{aligned}\hat{Z}_{tw}(0) &= (iw\hat{X}_w\tau)_w = (iw\hat{X}_w\tilde{\phi}(0))_w = \left(\hat{Y}_z\phi(0)\frac{dz}{dw}\frac{dw}{dz}\right)_w\\ &= (\hat{Y}_z\phi(0))_w = (\hat{Y}_z\phi(0))_z\frac{dz}{dw}.\end{aligned}$$

Then

$$\hat{Z}_{tw}(0)\cdot\hat{Z}_{tw}(0) = (\hat{Y}_z\phi(0))_z\cdot(\hat{Y}_z\phi(0))_z\left(\frac{dz}{dw}\right)^2 = \hat{Y}_{zz}\cdot\hat{Y}_{zz}\phi(0)^2\left(\frac{dz}{dw}\right)^2.$$

Thus,

$$w\hat{Z}_{tw}(0)\cdot\hat{Z}_{tw}(0)\tilde{\phi}_t(w)\,dw = \phi(0)^2\hat{Y}_{zz}\cdot\hat{Y}_{zz}\left(w\frac{dz}{dw}\right)\phi_t\,d\theta$$

proving Proposition 8.8 for the second term in (8.34). The expression for the first term follows similarly. □

Given that

$$w\frac{dz}{dw} = \left(\frac{1+z^2}{2i}\right)$$

(formulae (8.51) and (8.52)) we obtain another formula for the fourth derivative, namely:

$$\begin{aligned}\frac{d^4E}{dt^4} &= 6\,\mathrm{Im}\int_{S^1}(1+z^2)\hat{Z}_{ttz}(0)\cdot[\hat{Z}_{tz}(0)\tau + \hat{Y}_z\phi_t(0)]\,d\theta\\ &\quad + 6\,\mathrm{Im}\int_{S^1}(1+z^2)\hat{Y}_{zz}\cdot\hat{Y}_{zz}\tau^2\phi_t(0)\,d\theta.\end{aligned} \tag{8.81}$$

In order to show that (8.80) can be made negative, while all lower order derivatives vanish, we need:

Proposition 8.9 *Referring to formula* (8.74) ξ *and* ρ *can be chosen as* C^∞ *smooth functions, so that*

(a) $\xi(x) = \tau\sigma_x$, *where* $\tau = \sum\tau_j$, $\tau_j = \beta_j/z^j$, j *odd* $\beta_j \in \mathbb{R}$,

and

(b) $\phi_t(0) = \alpha/z^{2n} + \sum\alpha_j/z^{2i+1}$, $\alpha, \alpha_j \in \mathbb{R}$, $j < n$,

where each sum above is finite, i.e. $\tau, \phi_t(0)$, *are meromorphic functions on* $\mathcal{H}$ *(and consequently on* $\overline{B}$*).*

Proof From Lemma 8.4, we have

$$\phi(t)(x) = [\xi(x) + t\rho(x)]/[\sigma_x + t\xi_x + (t^2/2)\rho_x]. \tag{8.82}$$

Now choose

$$\xi(x) := \tau\sigma_x,$$

where τ is a meromorphic function on $\overline{B}$, real on S^1, such that $\hat{Z}_w(0)\tau$ is holomorphic. This choice assures that σ_t is monotonic for small $t > 0$, and that $\phi(0) = \tau$ is a generator of a forced Jacobi field.

An easy calculation shows that

$$\phi_t(0) = \rho/\sigma_x - \xi\xi_x/\sigma_x^2 = \rho/\sigma_x - (\tau\tau_x + \tau^2(\sigma_{xx}/\sigma_x)). \tag{8.83}$$

Write $\sigma_x = ax^n g(x)$, $g(0) = 1$ and $\sigma_{xx} = anx^{n-1} f(x)$, $f(0) = 1$, where f and g are (by Taylor's theorem) C^∞ smooth. Thus

$$\begin{aligned}\phi_t(0) &= \rho/ax^n g(x) - \tau\tau_x - (n\tau^2/x)(f/g)\\ &= \rho/ax^n g(x) - \tau\tau_x - (n\tau^2/x) + \left(\frac{g-f}{g}\right)(n\tau^2/x).\end{aligned}$$

Now

$$g - f = \sum_{j=1}^{2n} d_j x^j + h(x)x^{2n+1},$$

where $h \in C^{r-(2n+2)}$. Thus $h \in C^{r-3n}$. Let

$$\rho_1 := -\{h(x)x^{2n+1}\}ax^n(n\tau^2/x)$$

and $\rho = \rho_1' + \rho_1$.

Now

$$\begin{aligned}\frac{\rho_1'}{ax^n g} + \frac{(\sum_{j=1}^{2n} d_j x^j)n\tau^2/x}{g} &= \frac{\sum_{j=1}^{2n} d_j'/x^j}{g} + \frac{\rho_1'}{ax^n g}\\ &= \frac{\sum_{j=1}^{2n} d_j'/x^j}{g} - d_{2n}'/x^{2n} + d_{2n}'/x^{2n} + \rho_1'/ax^n g\\ &= \frac{\sum_{j=1}^{2n-1} d_j''/x^j + h_2(x)}{g} + d_{2n}'/x^{2n} + \rho_1'/ax^n g.\end{aligned}$$

Choose $\rho_1' = \rho_2' + \rho_2$, $\rho_2 = -ax^n h_2(x)$. Continue in this manner. Then

$$\phi_t(0) = -\tau\tau_x - n\tau^2/x + \cdots \tag{8.84}$$

where $+\cdots$ are meromorphic functions, real on S^1, with poles of lower order. □

Remark 8.6 In order to ensure monotonicity for the variations σ_t we need to account for other possible zeros of σ_x. In order to adjust for this we use Wienholtz's trick of adding to the definition of τ terms τ^* of the form $\alpha/z - z_x$, $\alpha > 0$, where z^* is a point in $\mathbb{R}$ where $\sigma_x(z^*) = 0$. These terms will not, as in the case of the third derivative, affect the final calculation of the fourth derivative since $n \geq 2$. We shall later discuss the monotonicity question.

We are now ready to apply (8.84) and the fact that $\phi(0) = \tau$ to the formula for the fourth derivative (8.81).

Theorem 8.5 *Suppose* $3n \leq 2m - 2 < 4(n-1)$. *Then*

$$\frac{d^4 E}{dt^4}(0) < 0$$

for an appropriate choice of τ.

Proof The trick will be to show that the methodology developed in Dierkes, Hildebrandt and Tromba [1] also applies to this case. Here we chose $\tau := \beta/z^{n-1} + \rho/z^n$ when $2m - 2 = 3(n-1) + r$, $r < (n-1)$.

Consider the expression

$$\hat{Z}_{tz}(0)\tau + \hat{Y}_z\phi_t(0). \tag{8.85}$$

Both the first and third complex components have poles, yet its real part $\hat{Z}_{tx}(0)\tau + Y_x\phi_t(0)$ has no pole. The last complex component may have poles of the form $\sum \mu_j R_j/z^{\ell_j}$, $\ell_j > 0$, $j \leq m+2$, $s \leq 2n - m$ where necessarily $\operatorname{Re} R_j = 0$ and the first complex component has pole terms of the form $\sum^n v_j A_j/z^{k_j}$, $j \leq n+1$, $k_j > 0$, and no pole forms on A_{n+2}, since in the Taylor expansion A_{n+2} has an initial exponent $2n+1$.

Now $t \to \hat{Z}(t)$ is sufficiently smooth as is

$$x \mapsto Z(t)(x) = X(\varphi_t(x)) = \gamma \circ \sigma_t(\lambda).$$

Therefore

$$(1+z^2)\left\{\hat{Z}_{tz}(0)\tau + \hat{Y}_z\phi_t(0) - \left(\sum v_j A_j/z^{k_j}, \sum \mu_i R_i/z^{\ell_i}\right)\right\}$$

is a *global* meromorphic function on B and $\mathcal{H}$, yet its real part $\hat{Z}_{tx}(0)\tau + \hat{Y}_x\phi_t(0)$ is continuous. Thus from Chaps. 2 and 3

$$\hat{Z}_{ttz}(0) = \left\{(1+z^2)\left[\hat{Z}_{tz}(0)\tau + \hat{Y}_z\phi_t(0) - \left(\sum v_j A_j/z^{k_j}, \sum \mu_i R_i/z^{\ell_i}\right)\right]\right\}_z.$$

This remarkable formula shows that, unlike removing poles as in the interior case, the smoothness of the contour Γ implies that the only poles that form in (8.85) are purely imaginary, allowing us to make the computations exactly as in the interior case, again using the fact that $A_j \cdot A_k = 0$, $1 \leq j, k \leq 2m - 2n$.

We have $\tau = \epsilon\beta/z^{n-1} + \mu/z^r$ and from (8.76)

$$-\phi_t(0) = (n - r + 1)\epsilon\beta^2\mu/z^{n+r} + \epsilon^2/z^{2n-1} + \cdots.$$

Consider the first term in (8.81). Noting that, as in the third derivative, the first complex components play no role, we obtain

$$(1+z^2)\hat{Z}_{tt}(0)\cdot[\hat{Z}_{tt}(0)\tau+\hat{Y}_t\phi_t(0)] = 2\epsilon^3\beta^3\mu(m-n-r)(m-n)^2R_m^2/z+\cdots. \tag{8.86}$$

Given that $1+z^2=2iw$ and $\phi=iw\tilde{\phi}\frac{dz}{dw}$, it follows that the left-hand side of (8.86) has a zero of order at least two at $w=0$.

Set $C_1 := 2\epsilon^3\beta^3\mu(m-n-r)(m-n)^2R_m^2/z$.

Now

$$\begin{aligned}
&6\,\mathrm{Im}\int(1+z^2)\hat{Z}_{tt}(0)\cdot[\hat{Z}_{tz}(0)\tau+\hat{Y}_z\phi_t(0)]\,d\theta\\
&=6\,\mathrm{Im}\int\left\{(1+z^2)\hat{Z}_{tt}(0)\cdot[\hat{Z}_{tz}(0)\tau+\hat{Y}_z\phi_t(0)]-C_1\right\}\frac{dw}{iw}.
\end{aligned}$$

The integrand is continuous and we can apply, as before, the Cauchy residue theorem to conclude that the first term of (8.81) equals

$$\begin{aligned}
&-12\pi\epsilon^3\beta^3\mu(m-n-r)(m-n)^2\,\mathrm{Im}\,\mathrm{Res}\left(i\frac{w+1}{w-1}R_m^2\right)_{w=0}\\
&=12\pi\epsilon^3\beta^3\mu(m-n-r)(m-n)^2R_m^2+\cdots
\end{aligned} \tag{8.87}$$

where $+\cdots$ means higher terms in ϵ. One sees that

$$\hat{Y}_{zz}\cdot\hat{Y}_{zz}=(m-n)^2R_m^2z^{2m-2}. \tag{8.88}$$

Then from Proposition 8.2, it follows that in the second term of (8.81) no pole is attached to any R_j if $\mathrm{Re}\,R_j\neq 0$. Then, by exactly the same reasoning, the second term of (8.81) is equal to

$$-6\pi\epsilon^3\beta^3\mu(n-r+3)(m-n)^2R_m^2+\cdots.$$

However, $2m-2=3(n-1)+r$, whence

$$2(m-n-r)=2m-2-2(n-1)-2r=(n-1-r),$$

yielding that (8.87) equals

$$6\pi\epsilon^3\beta^3\mu(n-1-r)(m-n)^2+\cdots$$

and thus the sum of the two terms of (8.81) equals $-24\pi\epsilon^3\beta^3\mu r(m-n)^2$.

Finally, we determine that

$$\frac{d^4E}{dt^4}(0)=-24\pi\epsilon^3\beta^3\mu r(m-n)^2R_m^2+\cdots \tag{8.89}$$

from which it immediately follows that if we can choose μ negative, then since $R_m^2<0$ the derivative (8.79) can be made negative for sufficiently small ϵ and we have proved the theorem. □

Now we have:

Proposition 8.10 *For* $r < (n-1)$, $\beta > 0$, *and* $\mu < 0$, *the variation* σ_t *can be made monotonic.*

Proof We have

$$\sigma_t(x) = \sigma + t\tau\sigma_x + (t^2/2)\rho + t\tau_*\sigma_x.$$

For ease of exposition let us assume that $\hat{X}$ has only one other boundary branch point at $z^* \in R$. We take

$$\tau = \epsilon\beta/z^{n-1} + \mu/z^r + \tau_*$$

and $\tau_* = v/(z - z^*)$; $v > 0$. Write $\sigma_x = ax^n + O(|x^{n+1}|)$, $a > 0$ and

$$\frac{d}{dx}\sigma_t = \sigma_x + t(\tau\sigma_x)_x + (t^2/2)\rho_x + t(\tau_*\sigma_x)_x. \tag{8.90}$$

Near $x = 0$ (8.90) equals

$$\sigma_x + at\beta\epsilon + t\beta O(|x|) + ta\mu x^{n-r-1} + t\mu O(|x|^{n-r})$$
$$+ t(t/2)\rho_x + tC_1 O(|x|^{n-1}). \tag{8.91}$$

Assume μ and β are fixed, say $\beta = 1$, $\mu = -1$. Choose $\epsilon > 0$ fixed so that the fourth derivative is negative. Then we see that (8.90) equals

$$\sigma_x + at(\epsilon - x^{n-r-1}) + tO(|x|) + t(t/2\rho_x). \tag{8.92}$$

Pick an interval of radius $\epsilon_1 < \epsilon/2 \ll 1$ and $t > 0$ small enough so that $O(|x|) + \frac{1}{2}|t \cdot \rho_x| < \epsilon/4$. Then on $[-\epsilon_1, \epsilon_1]$ we have (8.92), and therefore (8.90) is strictly positive.

Now denote by ℓ the order of the branch point z^*. Near $z = z^*$, $(\tau_*\sigma_x)_x = vC_2(z - z^*)^{\ell-2} + O(|z - z^x|^{\ell-1})$, $C_2 > 0$, $\sigma_x = C_3(z - z^*)^{\ell}$, $(C_3 > 0) + O(|z - z^*|^{\ell+1}) + (\tau\sigma_x)_x = tC_4(z - z^*)^{\ell-1} + tO(|z - z^*|^{\ell})$, and $(t^2/2)\rho_x = t(t/2)\rho_x$.

Then near $z = z^*$

$$\frac{d}{dx}\sigma_t = vtC_2(z - z^*)^{\ell-2} + tO(|z - z^*|^{\ell-1}) + C_3(z - z^*)^{\ell}$$
$$+ O(|z - z^*|^{\ell+1}) + t(t/2)\rho_x. \tag{8.93}$$

Pick an interval $I := [z^* - \epsilon_2, z^* + \epsilon_2]$ so that for $z \in \mathbb{R}$ in this interval

(i) $C_3(z - z^*)^{\ell} + O(|z - z^*|^{\ell+1}) > 0$;
(ii) $vC_2(z - z^*)^{\ell-2} + O(|z - z^*|^{\ell-1}) > 0$

and $0 < t < (vC_2/4\kappa)\epsilon_2^{\ell-2}$, where $\kappa = \sup|\rho_x|$ on I.

Then for t sufficiently small $(d/dx)\sigma_t > 0$ on $[-\epsilon_1, \epsilon_1] \cup [z^* - \epsilon_2, z^* + \epsilon_2]$. Since $\sigma_x > 0$ on the complement of the union of these two intervals and S^1 is compact, we can choose t small enough so that σ_t is monotonic. This concludes Proposition 8.10. □

Thus we have shown

Theorem 8.6 *If $\hat{X}_0$ is a minimal surface with a boundary branch point of even order n and index m, where $2m-2<4(n-1)$, then $\hat{X}_0$ cannot be a minimum of either energy or area, i.e. there is a C^∞ surface $\hat{X}$ with less energy and area than that of $\hat{X}_0$. If $\hat{X}_0$ maps S^1 monotonically onto Γ, $\hat{X}$ can be chosen to map S^1 monotonically onto Γ.*

We would now like to move on to the cases $3n \le 2m-2<4n$ (here we are still in the non-exceptional situation where $m+1 \ne 2(n+1)$) and $4n<2m-2\le 5n$. For both these situations we need to consider the fifth derivative in the direction of forced Jacobi fields. In this case, for ease of exposition, we omit the mention of τ_*.

Proposition 8.11

$$\frac{d^5E}{dt^5}=\mathrm{Re}\sum_{j=1}^{9} I_j \tag{8.94}$$

with

$$\begin{aligned}
I_1 &:= 16\int_{S^1} w\hat{Z}_{tttw}\cdot\hat{Z}_{tw}\tilde{\phi}\,dw, & I_2 &:= 12\int_{S^1} w\hat{Z}_{ttw}\cdot\hat{Z}_{ttw}\tilde{\phi}\,dw,\\
I_3 &:= 4\int_{S^1} w\hat{Z}_{ttttw}\cdot\hat{Z}_{w}\tilde{\phi}\,dw, & I_4 &:= 16\int_{S^1} w\hat{Z}_{tttw}\cdot\hat{Z}_{w}\tilde{\phi}_t\,dw,\\
I_5 &:= 48\int_{S^1} w\hat{Z}_{ttw}\cdot\hat{Z}_{tw}\tilde{\phi}_t\,dw, & I_6 &:= 24\int_{S^1} w\hat{Z}_{ttw}\cdot\hat{Z}_{w}\tilde{\phi}_{tt}\,dw,\\
I_7 &:= 24\int_{S^1} w\hat{Z}_{tw}\cdot\hat{Z}_{tw}\tilde{\phi}_{tt}\,dw, & I_8 &:= \int_{S^1} w\hat{Z}_{tw}\cdot\hat{Z}_{w}\tilde{\phi}_{ttt}\,dw,\\
I_9 &:= 2\int_{S^1} w\hat{Z}_{w}\cdot\hat{Z}_{w}\tilde{\phi}_{tttt}\,dw.
\end{aligned}\tag{8.95}$$

$I_3(0)$ vanishes by Cauchy's theorem since both $\hat{Z}_{tttt}(0)_w$ and $w\hat{X}_w\tau$ are holomorphic provided that $\tau=\phi(0)$ is the generator of a forced Jacobi field at $\hat{X}$. Furthermore, $I_8(0)=0$ because of (8.81), and $\hat{X}_w\cdot\hat{X}_w=0$ implies $I_9(0)=0$.

Thus we obtain from Chaps. 2 and 3

Proposition 8.12 *Since $\hat{X}$ is a minimal surface we have*

$$\begin{aligned}
\frac{d^5E}{dt^5}(0) = {}& 16\,\mathrm{Re}\int_{S^1} \hat{Z}_{ttw}(0)\cdot[w\hat{Z}_{tw}(0)\tilde{\phi}(0)+w\hat{X}\tilde{\phi}_t(0)]\,dw\\
&+12\,\mathrm{Re}\int_{S^1} Z_{ttw}(0)\cdot[w\hat{Z}_{ttw}(0)\tilde{\phi}(0)\\
&+4w\hat{Z}_{tw}(0)\tilde{\phi}_t(0)+2w\hat{X}_w\tilde{\phi}_{tt}(0)]\,dw\\
&-24\,\mathrm{Re}\int_{S^1} w^3\hat{X}_{ww}\cdot\hat{X}_{ww}\tilde{\phi}(0)^2\tilde{\phi}_{tt}(0)\,dw
\end{aligned}\tag{8.96}$$

provided that $\phi(0)$ is the generator of an inner forced Jacobi field at $\hat{X}$.

Lemma 8.7 *If $f(w) := w\hat{Z}_{tw}(0)\tau + w\hat{X}_w\phi_t(0)$ is holomorphic, then*

$$\hat{Z}_{ttw}(0) = \{iw[iw\hat{X}_w\tau]_w\tilde{\phi}(0) + iw\hat{X}_w\tilde{\phi}_t(0)\}w, \tag{8.97}$$

and

$$\hat{Z}_{ttw}(0)\cdot\hat{X}_w = -\hat{Z}_{tw}(0)\cdot\hat{Z}_{tw}(0) = w^2\hat{X}_{ww}\cdot\hat{X}_{ww}\tau^2. \tag{8.98}$$

Proof The proofs of (8.97) and (8.98) are in Lemma 2.4 of Chap. 2. □

Since $2m-2 > 4n$, $\hat{Z}_{tz}(0)\tau + Y_z\phi_t(0)$ has no pole in the last complex component. Since $\hat{Z}_{tx}(0)\tau + Y_x\phi_t(0)$ has no pole, the only possible poles are imaginary; i.e. $\Sigma\nu_j A_j/z^{k_j}$, $\nu_j \in R$, $j \le n+1$ or $\Sigma\nu'_j A_1/z^{k_j} = (\Sigma\nu'_j/z^{k_j}, i\Sigma\nu'_j/z^{k_i})$. Thus $\nu'_j = 0$ and so $f(w)$ is holomorphic.

As a consequence if $f(w)$ is holomorphic, we have a formula for the fifth derivative, namely

Proposition 8.13 *If $f(w) := w\hat{Z}_{tw}(0)\tilde{\phi}(0) + w\hat{X}_w\tilde{\phi}_t(0)$ is holomorphic, then*

$$\begin{aligned}\frac{d^5E}{dt^5}(0) = 12\,\mathrm{Re}\int_{S^1}&[w\hat{Z}_{ttw}(0)\cdot\hat{Z}_{ttw}(0)\tilde{\phi}(0)\\&+4w\hat{Z}_{ttw}(0)\cdot\hat{Z}_{tw}(0)\tilde{\phi}_t(0)]\,dw.\end{aligned} \tag{8.99}$$

As we did with the fourth derivative, we can write (8.99) in the notation of the upper half-plane in order to aid in further computations.

Noting that

$$Z_{ttw} = Z_{ttz}\frac{dz}{dw}, \qquad Z_{tw} = Z_{tz}\frac{dz}{dw}, \qquad \tilde{\phi} = \frac{1}{iw}\frac{dw}{dz}\phi,$$

$$\tilde{\phi}_t = \frac{1}{iw}\frac{dw}{dz}\phi_t, \qquad w\frac{dz}{dw} = \left(\frac{1+z^2}{2i}\right), \qquad d\theta = dw/iw,$$

we obtain: if $f(w)$ above is holomorphic

$$\frac{d^5E}{dt^5}(0) = 6\,\mathrm{Im}\int_0^{2\pi}(1+z^2)Z_{ttz}(0)\cdot\big[Z_{tt}(0)\phi(0) + 4Z_{tz}(0)\phi_t(0)\big]\,d\theta. \tag{8.100}$$

We are now ready to compute this integral, or at least terms of a certain ϵ-order.

We first consider the situation $4n < 2m-2 \le 5n$. Then

$$2m-3 = 4n+r, \quad n > r > 0 \text{ odd}. \tag{8.101}$$

Here we set $\tau = \beta\epsilon/z^n + \mu/z^r$, $\tau = \phi(0)$. Then formula (8.84) applies to give us $\phi_t(0)$. Since $2m-2 \ge 4n+2$, $2m-3 \ge 4n+1$, and this implies that the $f(w)$ of Lemma 8.7 is in fact holomorphic.

Then, we have that the last complex component of

$$\hat{Z}_{tz}(0)\phi(0) + \hat{X}_z\phi_t(0) \tag{8.102}$$

equals

$$\begin{aligned}&\beta^2\epsilon^2(m-n)R_m z^{m-2n-1}\\&\quad+2\beta\mu\epsilon(m-n)R_m z^{m-n-r-1}\\&\quad+\mu^2(m-n)R_m z^{m-2r-1}\\&\quad+\text{higher order terms in powers of } z.\end{aligned}\tag{8.103}$$

Moreover, the last complex component of $\hat{Z}_{ttz}$ equals

$$\begin{aligned}&\beta^2\epsilon^2(m-2n-1)(m-n)R_m z^{m-2n-2}\\&\quad+2\epsilon(m-n)(m-n-r-1)R_m z^{m-n-r-2}\\&\quad+\mu^2(m-n)(m-2r-1)R_m z^{m-2r-2}\\&\quad+\text{higher order terms in powers of } z.\end{aligned}\tag{8.104}$$

The last complex component of $\hat{Z}_{ttz}(0)\cdot\hat{Z}_{ttz}(0)$ equals

$$\begin{aligned}&\beta^4\epsilon^4(m-2n-1)^2(m-n)^2R_m^2z^{2m-4n-4}\\&\quad+4\epsilon^4\mu\beta^2(m-n)^2(m-n-r-1)^2R_m^2z^{2m-2n-2r-4}\\&\quad+(m-n)^2(m-2r-1)^2R_m^2z^{2m-4r-4}\\&\quad+4\beta^3\epsilon^3\mu(m-2n-1)(m-n)^2(m-n-r-1)R_m^2z^{2m-3n-r-4}\\&\quad+2\beta^2\epsilon^3\mu^2(m-n)^2(m-2n-1)^2R_m^2z^{2m-2n-2r-4}\\&\quad+4\epsilon\mu^3\beta(m-n)^2(m-n-r-1)(m-2r-1)R_m^2z^{2m-n-3r-4}\\&\quad+\text{higher order terms in } z.\end{aligned}$$

Thus, the last complex component of $\hat{Z}_{ttz}(0)\cdot\hat{Z}_{tz}(0)\phi(0)$ equals

$$\begin{aligned}&\beta^4\mu\epsilon^4(m-n)^2(m-2n-1)\{(m-2n-1)+4(m-n-r-1)\}R_m/z\\&\quad+O(\epsilon^5)+\text{terms with no poles in } z,\end{aligned}\tag{8.105}$$

and the last complex component of $\hat{Z}_{ttz}(0)\cdot Z_{tz}(0)\phi_t(0)$ equals

$$-\beta^4\epsilon^4\mu(m-2n-1)(m-n)^2(n-r)R_m^2z^{2m-4n-r-4}+\cdots\tag{8.106}$$

where

$$2m-4n-r-4=-1.$$

Thus, noting again that the relation $A_i\cdot A_j=0$, $1\le i,j\le 2m-2n$ implies that the first complex components do not contribute to the ϵ^4 terms in the derivative, we have

$$\frac{d^5E}{dt^5}(0)=6\beta^4\epsilon^4\mu(m-n)^2\gamma(m-2n-1)\,\mathrm{Im}\int[1+z^2]\gamma\frac{R_m^2}{z}\,d\theta+O(\epsilon^5)\tag{8.107}$$

where

$$\gamma := [(m - 2n - 1) + 4(m - n - r - 1) - 4(n - r)]$$
$$= [5(m - 2n - 1)].$$

Again, using the fact that $1 + z^2 = 2iw$ we may apply the same analysis as in the fourth derivative to conclude that

$$\frac{d^5 E}{dt^5}(0) = -\xi \epsilon^4 \mathrm{Re}\, s \left[R_m^2 \frac{1}{w} \left(\frac{w-1}{w+1} \right) \right]_{w=0} + O(\epsilon^5) \tag{8.108}$$

where $\xi := 60\pi\beta^4\mu(m-n)^2(m-2n-1)^2 > 0$. Therefore, in the case $4n < 2m - 2 \leq 5n$, the fifth derivative of Dirichlet's energy is

$$\frac{d^5 E}{dt^5}(0) = \xi \epsilon^4 R_m^2 + O(\epsilon^5) \tag{8.109}$$

which (since $R_m^2 < 0$) can be made negative for small ϵ.

Noting that with $\mu > 0$ and with $\operatorname{sgn} \beta = \operatorname{sgn} b_2$ (cf. (8.73)), it follows as before that the variation σ_t is monotone if $b_2 = 0$ choose $\tau = \beta/z^n + \mu/r^n + \gamma/z^{n-1}$. Then the fifth derivative is unchanged but monotonicity is preserved. Thus we have proved:

Theorem 8.7 *If $\hat{X}_0$ is a minimal surface spanning a smooth contour with a boundary branch point of order n and index m, where $4n < 2m - 2 \leq 5n$, then $\hat{X}_0$ cannot be a minimum of either area or energy; i.e. there is a C^∞ surface $\hat{X}$ spanning Γ with less area or energy. If X_0 maps S^1 monotonically onto Γ, $\hat{X}$ will also map S^1 monotonically onto Γ.*

Next, we revisit the cases left over from considerations of the fourth derivative, namely

(A) $2m - 2 = 4n - 4$ or $m = 2n - 1$,
(B) $2m - 2 = 4n - 2$ or $m = 2n$.

In both cases $m \geq 2(n-1) + 1$, and so Lemma 8.4 applies and formulae (8.96) and (8.99) for the fifth derivative hold.

We can write

$$2m - 3 = 2(n-1) + 2(n-2) + 1 \quad \text{(Case A)}$$

or

$$2m - 3 = 2(n-1) + 2(n-3) + 3 \quad \text{(Case B)}.$$

Thus for both (A) and (B) we can write

$$2m - 3 = 2k_1 + 2k_2 + r,$$

$k_1 = (n-1)$, $k_2 = (n-2)$ and r is 1 or 3.

With these generators we will need to consider the first as well as the last complex components. We can consider both cases simultaneously by writing our generator τ as

$$\phi(0) = \tau := \beta\epsilon/z^{k_1} + \gamma/z^{k_2} + \rho/z^r,$$

$2k_1 + 2k_2 + r = 2m - 3, r > 0$, odd, $r \le k_2$. Then $\phi_t(0)$ is determined by (8.84).

We see that since in both cases (A) and (B), $4r \le 4k_2 < 4k_1 \le 2m - 2$, the fourth derivative as well as the second and third, with this choice of τ, vanish identically. In (A) in order to ensure that $r \le k_2$, we need to assume that $n > 2$. Thus, in the following considerations we omit the cases $n = 2, m = 3, m = 4$. However, the case $n = 2, m = 3$ cannot occur by Wienholtz's result since here $2m - 2 = 4 < 6 = 3n$. Hence we omit $n = 2, m = 4$, and we therefore (in (A)) assume that $n \ge 4$. In case (B), if $n = 4, m = 8$, and thus $r = (n-1)$, not $r < (n-1)$. Thus we leave consideration of this case until after of looking at case (B). With these exceptions we consider cases (A) and (B) simultaneously.

We first evaluate only the last complex components of $\hat{Z}_{tz}(0)$, $\hat{Z}_{ttz}(0)$, $\hat{Z}_{ttz}(0) \cdot \hat{Z}_{ttz}(0)\phi(0)$ and $\hat{Z}_{ttz}(0) \cdot \hat{Z}_{tz}(0)\phi_t(0)$.

Now thc last complex component of $\hat{Z}_{tz}(0)$ equals

$$\beta\epsilon(m - k_1)R_m z^{m-k_1-1} + \gamma(m - k_2)R_m z^{m-k_2-1} + \rho(m - r)R_m z^{m-r-1}. \quad (8.110)$$

The last complex component of $\hat{Z}_{tz}(0)\phi(0)$ equals

$$\begin{aligned}&\epsilon^2\beta^2(m - k_1)R_m z^{m-2k_1-1} + \beta\gamma\epsilon(2m - k_1 - k_2)R_m z^{m-k_1-k_2-1}\\ &+ \beta\epsilon\rho(2m - k_1 - r)R_m z^{m-k_1-r-1} + \gamma^2(m - k_2)R_m z^{m-2k_2-1}\\ &+ \rho^2(m - r)R_m z^{m-2r-1} + \gamma\rho(2m - k_2 - r)R_m z^{m-k_2-r-1}.\end{aligned} \quad (8.111)$$

Also we have, from (8.84)

$$\begin{aligned}-\phi_t(0) = &\beta^2\epsilon^2(n - k_1)z^{-(2k_1+1)} + \epsilon\beta\gamma(2n - k_1 - k_2)z^{-(k_1+k_2+1)}\\ &+ \beta\epsilon\rho(2n - k_1 - r)z^{-(k_1+r+1)} + \gamma^2(n - k_2)R_m z^{-(2k_2+1)} + \cdots\end{aligned} \quad (8.112)$$

(we ignore the term ρ^2).

Therefore, the last complex component of $\hat{Z}_{tz}(0)\phi(0) + X_z\phi_t(0)$ equals

$$\begin{aligned}&\beta^2\epsilon^2(m - n)R_m z^{m-2k_1-1} + 2\beta\gamma\epsilon(m - n)R_n z^{m-k_1-k_2-1}\\ &+ 2\beta\epsilon\rho(m - n)R_m z^{m-k_1-r-1} + \gamma^2(m - n)R_m z^{m-2k_2-1}\\ &+ 2\gamma\rho(m - n)R_m z^{m-k_2-r-1} + \cdots.\end{aligned} \quad (8.113)$$

Thus, the last complex component of $\hat{Z}_{ttz}(0)$ is

$$\begin{aligned}&\beta^2\epsilon^2(m - n)(m - 2k_1 - 1)R_m z^{m-2k_1-2}\\ &+ 2\beta\gamma\epsilon(m - n)(m - k_1 - k_2 - 1)R_m z^{m-k_1-k_2-2}\\ &+ 2\beta\epsilon\rho(m - n)(m - k_1 - r - 1)R_m z^{m-k_1-r-2}\end{aligned}$$

$$+\gamma^2(m-n)(m-2k_2-1)R_m z^{m-2k_2-2}$$
$$+2\gamma\rho(m-n)(m-k_2-r-1)R_m z^{m-k_2-r-2}+\cdots. \qquad (8.114)$$

To make the next computations somewhat simpler, we ignore terms involving ρ^2, γ^4, ρ^4. Adopting this strategy, we see that the last complex component of $\hat{Z}_{ttz}(0)\cdot\hat{Z}_{ttz}(0)$ equals

$$\begin{aligned}
&\beta^4\epsilon^4(m-n)^2(m-2k_1-1)^2R_m^2 z^{2m-4k_1-4}\\
&\quad+4\beta^2\epsilon^2\gamma^2(m-n)^2(m-k_1-k_2-1)^2R_m^2 z^{2m-2k_1-2k_2-4}\\
&\quad+4\beta^2\epsilon^2\gamma\rho(m-n)^2(m-k_1-k_2-1)(m-k_1-r-1)R_m^2 z^{2m-2k_1-k_2-r-4}\\
&\quad+4\beta\epsilon\gamma^2\rho(m-n)^2(m-k_1-k_2-1)(m-k_2-r-1)R_m^2 z^{2m-2k_2-k_1-r-4}.
\end{aligned} \qquad (8.115)$$

Now computing only the $\epsilon^2\beta^2\gamma^2\rho$ terms of the last complex component of $\hat{Z}_{ttz}(0)\cdot\hat{Z}_{ttz}(0)\phi(0)$, we obtain

$$\begin{aligned}
&4\beta^2\epsilon^2\gamma^2\rho\{(m-n)^2(m-k_1-k_2-1)^2\\
&\quad+(m-n)^2(m-k_1-k_2-1)(m-k_1-r-1)\\
&\quad+(m-n)^2(m-k_2-k_1-1)(m-k_2-r-1)\}R_m/z.
\end{aligned} \qquad (8.116)$$

Finally the last complex component of $-\hat{Z}_{ttz}(0)\cdot\hat{Z}_{tz}(0)\phi_t(0)$ (only the $\beta^2\gamma^2\epsilon^2\rho$ terms) equals

$$\begin{aligned}
&\beta^2\epsilon^2\gamma^2\rho\{(m-n)(m-k_2-r-1)(m-k_2)(n-k_1)\\
&\quad+(m-n)(m-2k_2-1)(m-r)(n-k_1)\\
&\quad+2(m-n)(m-k_2-r-1)(m-k_1)(2n-k_1-k_2)\\
&\quad+2(m-n)(m-k_1-r-1)(m-k_2)(2n-k_1-k_2)\\
&\quad+2(m-n)(m-k_1-k_2-1)(m-r)(2n-k_1-k_2)\\
&\quad+(m-n)(m-2k_2-1)(m-k_1)(2n-k_2-r)\\
&\quad+2(m-n)(m-k_1-k_2-1)(m-k_2)(2n-k_2-r)\\
&\quad+2(m-n)(m-k_2-r-1)(m-k_1)(n-k_2)\\
&\quad+(m-n)(m-2k_1-1)(m-r)(n-k_2)\}(R_m/z).
\end{aligned} \qquad (8.117)$$

In these formulae we need only substitute the relevant values for case (A) or case (B). We begin with case (A). In this case we have the following values: $k_1=(n-1)$, $k_2=(n-2)$, $2m-2=4(n-1)$, $r=1$, $(m-k_1-k_2-1)=1$, $(m-k_2-r-1)=(n-1)$, $(m-2k_1-1)=0$, $(m-n)=(n-1)$, $(m-k_2)=(n+1)$, $(m-k_1)=n$, $(n-k_1)=1$, $(n-k_2)=2$, $(2n-k_1-k_2)=3$, $(2n-k_2-r)=(n+1)$, $(m-k_1-r-1)=(n-2)$. Substituting these values into (8.117) we see that the last complex component of $\hat{Z}_{ttz}(0)\cdot\hat{Z}_{ttz}(0)\phi(0)$ equals

$$4(n-1)\{2n^2-4n+2\}\beta^2\epsilon^2\gamma^2\rho R_m^2/z+\cdots. \qquad (8.118)$$

On the other hand the last complex component of $\hat{Z}_{ttz}(0) \cdot \hat{Z}_{tz}(0)\phi_t(0)$ equals

$$(n-1)\{22n^2 + 6n - 28\}\beta^2\epsilon^2\gamma^2\rho R_m^2/z. \tag{8.119}$$

Therefore the leading term of the last complex component of

$$\hat{Z}_{ttz}(0) \cdot \hat{Z}_{ttz}(0)\phi(0) + 4\hat{Z}_{ttz}(0) \cdot \hat{Z}_{tz}(0)\phi_t(0)$$

equals

$$4(n-1)\{-20n^2 - 10n + 30\}\beta^2\epsilon^2\gamma^2\rho R_m^2/z. \tag{8.120}$$

As mentioned, with these generators, we must consider the contribution of the first complex components arising from the term $\hat{Z}_{ttz}(0) \cdot \hat{Z}_{tz}(0)$. Calculating this we get a final result of $2(n-1)\{-24n^2+12n+3\}$ and $-24n^2+12n+3$ is clearly negative for $n > 1$.

Let $\xi := 4(n-1)\{-24n^2 + 12n + 3\}$ and take $\beta > 0$ and $\rho < 0$. Then, as in the fourth derivative the variation σ_t is monotonic. The same analysis as before yields that the fifth derivative of Dirichlet's energy is

$$\frac{d^5 E}{dt^5}(0) = -2\pi\xi\beta^2\epsilon^2\rho\gamma^2 R_m^2 + O(\epsilon^3). \tag{8.121}$$

Since $R_m^2 < 0$ and $\rho < 0$, the derivative is negative for sufficiently small ϵ. Therefore, with the exception of $n = 2$, $m = 4$, case (A) is proved.

What about case (B), $n \neq 4$? Then we have: $m = 2n$, $k_1 = (n-1)$, $k_2 = n-2$, $r = 3$ with $n - 1 > 3$, and $(m - k_1 - k_2 - 1) = 2$, $(m - k_2 - r - 1) = (n-2)$, $(m - 2k_2 - 1) = 3$, $(m - 2k_1 - 1) = 1$, $(m - n) = n$, $(m - k_2) = (n+2)$, $(m - k_1) = (n+1)$, $(n - k_1) = 1$, $(n - k_2) = 2$, $(2n - k_1 - k_2) = 3$, $(2n - k_2 - r) = (n-1)$, $(m - k_1 - r - 1) = (n-3)$.

Then $\hat{Z}_{ttz}(0) \cdot \hat{Z}_{ttz}(0)\phi(0)$ has the leading term of the last complex component equal to

$$\beta^2\epsilon^2\gamma^2\rho\{4n(4n^2 - 4n)\}R_m^2/z \tag{8.122}$$

and the leading term of the last complex component of $\hat{Z}_{ttz}(0) \cdot \hat{Z}_{tz}(0)\phi_t(0)$ equals

$$-n\{23n^2 + 10n - 108\}\beta^2\epsilon^2\gamma^2\rho R_m^2/z. \tag{8.123}$$

Thus, the leading term of the integrand arising from the last complex component of

$$\hat{Z}_{ttz}(0) \cdot \hat{Z}_{ttz}(0)\phi(0) + 4\hat{Z}_{ttz}(0) \cdot \hat{Z}_{tz}(0)\phi_t(0)$$

is

$$4n\{-19n^2 - 14n + 108\}\beta^2\epsilon^2\gamma^2\rho R_m^2/z. \tag{8.124}$$

Calculating the contribution from the first complex components we obtain a final answer of $2n\{-24n^2 + 7n + 85\}$. Now $\xi := 2n\{-24n^2 + 7n + 85\} < 0$ for $n > 4$.

Take $\beta > 0$, $\rho < 0$, ensuring the monotonicity of σ_t. Then

$$\frac{d^5 E}{dt^5}(0) = 2\pi \xi \beta^2 \epsilon^2 \gamma^2 \rho R_m^2 + O(\epsilon^3) < 0 \tag{8.125}$$

for ϵ sufficiently small.

We have one case remaining, namely $n = 4$, $2m - 2 = 4n - 2$, $m = 8$. Here $2m - 3 = 13 = 4 \cdot 3 + 1$. Take $k = 3$, $r = 1$ and generator

$$\tau : \beta\epsilon/z^k + \mu/z^r. \tag{8.126}$$

Then an explicit calculation shows that

$$\frac{d^5 E}{dt^5}(0) = 2\pi \xi \beta^4 \epsilon^3 \mu R_m^2 + O(\epsilon^4) \tag{8.127}$$

where $\xi := (16)$.

Then, again, for $\epsilon > 0$ sufficiently small,

$$\frac{d^5 E}{dt^5}(0) < 0.$$

Thus, we have proved Theorem 8.2 in cases (A) and (B) with the exception of $n = 2$, $m = 4$.

We would now like to discuss the case $5n < 2m - 2 \leq 6n(2m + 2) < 6(n + 1)$. We first need a revision of Proposition 8.9, which is proved in the same way.

Proposition 8.14 *Consider a variation of σ*

$$\sigma_t(x) := \sigma(x) + t\xi(x) + (t^2/2)\rho(x) + (t^3/3)\eta(x). \tag{8.128}$$

Then ρ can be chosen as in Proposition 8.8 and η can be chosen so that

$$\phi_{tt}(0) = 2\tau(\tau_x + n\tau/x)^2 + \cdots \tag{8.129}$$

where $+\cdots$ are meromorphic functions on $\mathcal{H}$ with poles at $z = 0$ of lower order. We start by assuming $2m - 4 < 6n - 2$ and taking a generator of the form $c\epsilon^2/z^n + \gamma\epsilon/z^{k_1} + \delta/z^{k_2}$, $k_1 = n - 1$, $k_2 < k_1$. Then the derivatives $D_t^\beta \phi(0)$ do not contribute to the sixth derivative. We then have a sixth derivative given by

$$\begin{aligned}\frac{d^6 E}{dt^6}(0) = 40\,\mathrm{Re} \int \hat{Z}_{tttw}(0)& \\ \cdot \Big\{ w\hat{Z}_{ttw}(0)\tilde{\phi}(0) + 2w\hat{Z}_{tw}(0)\tilde{\phi}_t(0) + w\hat{Z}_w(0)\tilde{\phi}_{tt}(0) \Big\}\, dw& \\ + 60\,\mathrm{Re} \int \hat{Z}_{ttw}(0) \cdot \Big\{ w\hat{Z}_{ttw}(0)\tilde{\phi}_t(0) + 2w\hat{Z}_{tw}(0)\tilde{\phi}_{tt}(0) \Big\}\, dw.&\end{aligned} \tag{8.130}$$

Again, changing variables to the upper half-plane, we have

$$\frac{d^6E}{dt^6}(0) = 20\,\mathrm{Im}\int (1+z^2)\hat{Z}_{tttz}(0)$$
$$\cdot\left\{\hat{Z}_{ttz}(0)\phi(0) + z\hat{Z}_{tz}(0)\phi_t(0) + \hat{Z}_z(0)\phi_{tt}(0)\right\} d\theta$$
$$+ 30\,\mathrm{Im}\int (1+z^2)\hat{Z}_{ttz}(0)\cdot\left\{\hat{Z}_{ttz}(0)\phi_t(0) + 2\hat{Z}_{tz}(0)\phi_{tt}(0)\right\} d\theta. \quad (8.131)$$

Lemma 8.8

$$z \mapsto \mathrm{Re}\{\hat{Z}_{ttz}(0)\phi(0) + z\hat{Z}_{tz}(0)\phi_t(0) + \hat{Z}_z(0)\phi_{tt}(0)\} \quad (8.132)$$

is smooth.

Proof We have

$$Z'(t) = X(\varphi_t(x))_x\phi,$$
$$Z''(t) = [X(\varphi_t(x))_x\phi]\phi + X(\varphi_t(x))_x\phi_t.$$

At $t = 0$

$$Z''(0) = (X_x\phi)_x\phi + X_x\phi_t(0).$$

Similarly

$$Z'''(0) = [(X_x\phi)_x\phi + X_x\phi]_x\phi + 2[X_x\phi]_x\phi_t(0) + X_x\phi_{tt}(0)$$
$$= Z''(0)_x\phi + 2Z'(0)_x\phi_t(0) + X_x\phi_{tt}(0).$$

Since $x \mapsto Z'''(t)$ is smooth

$$x \mapsto Z''(0)_x\phi + 2Z'(0)_x\phi_t(0) + Z_x(0)\phi_{tt}(0) =: \Delta$$

is smooth. But

$$2\Delta = \mathrm{Re}\{\hat{Z}_{ttz}(0)\phi(0) + 2\hat{Z}_{tz}(0)\phi_t(0) + \hat{Z}_t(0)\phi_{tt}(0)\}$$

proving the lemma. □

We now calculate the sixth derivative as we have calculated the fourth, noting that any poles of

$$\hat{Z}_{ttz}(0)\phi(0) + 2\hat{Z}_{tz}(0)\phi_t(0) + \hat{Z}_z(0)\phi_{tt}(0)$$

are purely imaginary.

In all subsequent equalities we first consider only the last complex component. Since we are interested in the $\gamma\delta$-terms we will **ignore** the γ^2 and δ^2-terms in our calculations. We then have

$$\hat{Z}_{tz}(0) = \epsilon^2(m-n)R_m z^{m-n-1} + \epsilon\gamma(m-k_1)R_m z^{m-k_1-1}$$
$$+ \delta(m-k_2)R_m z^{m-k_2-1} \quad (8.133)$$

and

$$-\phi_t(0) = \gamma\epsilon^3(n-k_1)R_m z^{-(n+k_1+1)} + \delta\epsilon^2(n-k_2)R_m z^{-(n+k_2+1)} \\ + \delta\gamma\epsilon(2n-k_1-k_2)R_m z^{-(k_1+k_2+1)}. \tag{8.134}$$

Furthermore

$$\hat{Z}_{tz}(0)\phi + \hat{Z}_z(0)\phi_t(0) \\ = \epsilon^4(m-n)R_m z^{m-2n-1} + 2\epsilon^3\gamma(m-n)R_m z^{m-n-k_1-1} \\ + 2\epsilon^2\delta(m-n)R_m z^{m-k_2-n-1} + 2\epsilon\gamma\delta(m-n)R_m z^{m-k_1-k_2-1}. \tag{8.135}$$

Thus, we have

$$\hat{Z}_{ttz}(0) = \epsilon^4(m-n)(m-2n-1)R_m z^{m-2n-2} \\ + 2\gamma\epsilon^3(m-n)(m-n-k_1-1)R_m z^{m-n-k_1-2} \\ + 2\delta\epsilon^2(m-n)(m-n-k_2-1)R_m z^{m-n-k_2-2} \\ + 2\gamma\delta\epsilon(m-n)(m-k_1-k_2-1)R_m z^{m-k_1-k_2-2} \tag{8.136}$$

and hence

$$\hat{Z}_{ttz}(0)\phi = \epsilon^6(m-n)(m-2n-1)R_m z^{m-3n-2} \\ + 2\gamma\epsilon^5(m-n)(m-n-k_1-1)R_m z^{m-2n-k_2-2} \\ + 2\delta\epsilon^4(m-n)(m-n-k_2-1)R_m z^{m-n-k_1-k_2-2} \\ + \gamma\epsilon^5(m-n)(m-2n-1)R_m z^{m-2n-k_1-2} \\ + 2\gamma\delta\epsilon^3(m-n)(m-n-k_2-1)R_m z^{m-n-k_1-k_2-2} \\ + \delta\epsilon^4(m-n)(m-2n-1)R_m z^{m-2n-k_2-2} \\ + 2\gamma\delta\epsilon^3(m-n)(m-n-k_1-1)R_m z^{m-n-k_1-k_2-2} \\ + \delta\epsilon^4(m-n)(m-2n-1)R_m z^{m-2n-k_2-2} \\ + 2\gamma\delta\epsilon^3(m-n)(m-n-k_1-1)R_m z^{m-n-k_1-k_2-2} \tag{8.137}$$

and

$$-2\hat{Z}_{tz}(0)\phi_t(0) = 2\epsilon^5\gamma(m-n)(n-k_1)R_m z^{m-2n-k_1-2} \\ + 2\epsilon^4\delta(m-n)(n-k_2)R_m z^{m-2n-k_2-2} \\ + 2\epsilon^3\gamma\delta(m-k_2)(n-k_1)R_m z^{m-n-k_1-k_2-2} \\ + 2\epsilon^3\gamma\delta(m-k_1)(n-k_2)R_m z^{m-n-k_1-k_2-2} \\ + 2\epsilon^3\gamma\delta(m-n)(2n-k_1-k_2)R_m z^{m-n-k_1-k_2-2}. \tag{8.138}$$

Additionally, since

$$\phi_{tt}(0) = 4\gamma\delta\epsilon^3(n-k_1)(n-k_2)\cdot z^{(n-k_1+k_2+2)} + \cdots, \tag{8.139}$$

then

$$\hat{Z}_z(0)\phi_{tt}(0) = 4\epsilon^3\gamma\delta(n-k_1)(n-k_2)R_m z^{m-(n+k_1+k_2+2)}. \tag{8.140}$$

We are setting $2m-4=4n+k_1+k_2$, or

$$m=2n+[(k_1+k_2)/2]+2. \tag{8.141}$$

Keeping this in mind, we see that

$$\hat{Z}_{ttz}(0)\phi+2\hat{Z}_{tz}(0)\phi_t(0)+\hat{Z}_z(0)\phi_{tt}(0) \tag{8.142}$$

has only a γ-linear pole or a pole containing no γ or δ terms.

Using the fact that in evaluating the derivative, we multiply the zeros of $\hat{Z}_{tt}(0)$ by the poles of (8.142) to obtain all $\delta\gamma$-terms, we must then multiply the δ linear zeros of the z-derivative of (8.142) by the γ poles of (8.142) and the product of the $\gamma\delta$ zeros of the derivative of (8.142) by the poles of (8.142) contain no δ or ρ terms.

Let us first simplify the $\gamma\delta$ zeros of (8.142). From (8.137)–(8.139) we have the $\gamma\delta$ zeros of (8.142) equal to the product of $z^{m-n-k_1-k_2-2}$ and

$$\begin{aligned}
&2\gamma\delta\epsilon^3(m-n)(n-k_1-k_2-1)R_m+2\gamma\delta\epsilon^3(m-n)(m-n-k_2-1)R_m\\
&\quad+2\gamma\delta\epsilon^3(m-n)(m-n-k_1-1)R_m-2\gamma\delta\epsilon^3(m-k_1)(n-k_2)R_m\\
&\quad-2\gamma\delta\epsilon^3(m-k_2)(n-k_1)R_m-2\gamma\delta\epsilon^3(m-n)(2n-k_1-k_2)R_m\\
&\quad-2\gamma\delta\epsilon^3(n-k_1)(n-k_2)R_m-2\gamma\delta\epsilon^3(n-k_1)(n-k_2)R_m.
\end{aligned} \tag{8.143}$$

Adding these in steps, we first obtain

$$\begin{aligned}
&2\gamma\delta\epsilon^3(m-n)(m-2n-1)R_m+2\gamma\delta\epsilon^3(m-n)(m-n-k_2-1)R_m\\
&\qquad+2\gamma\delta\epsilon^3(m-n)(m-n-k_1-1)R_m-2\gamma\delta\epsilon^3(m-n)(n-k_2)R_m\\
&\qquad-2\gamma\delta\epsilon^3(m-n)(n-k_1)R_m\\
&\quad=6\gamma\delta\epsilon^3(m-n)(m-2n-1)R_m.
\end{aligned} \tag{8.144}$$

The pole term of (8.142) containing neither γ nor δ-terms is

$$\epsilon^6(m-n)(m-2n-1)R_m z^{m-3n-2}. \tag{8.145}$$

Thus, we have a contribution to the first term of the integrand of the sixth derivative arising from the product of the derivative of the $\gamma\delta$ zeros of (8.142) and the poles of (8.142) equalling

$$\begin{aligned}
&6\gamma\delta\epsilon^9(m-n)^2(m-2n-1)^2(m-n-k_1-k_2-2)R_m^2z^{-1}\\
&\quad=3\gamma\delta\epsilon^9(m-n)^2(m-2n-1)^2(2m-2n-2(k_1-k_2)-4)R_m^2z^{-1}\\
&\quad=3\gamma\delta\epsilon^9(m-n)^2(m-2n-1)^2(n-k_2+1)R_m^2z^{-1}
\end{aligned} \tag{8.146}$$

(since $2m-4=4n+k_1+k_2$, and $k_1=(n-1)$).

What about the δ zero of the derivative of (8.142) multiplied by the γ pole of (8.142)? From (8.137) and (8.138) the δ zero of (8.142) is

$$\begin{aligned}
&2\delta\epsilon^4(m-n)(m-n-k_2-1)R_m z^{m-2n-k_2-2}\\
&\quad+\delta\epsilon^4(m-n)(m-2n-1)R_m z^{m-2n-k_2-2}\\
&\quad-2\delta\epsilon^4(m-n)(n-k_2)R_m z^{m-2n-k_2-2}\\
&=3\delta\epsilon^4(m-n)(m-2n-1)R_m z^{m-2n-k_2-2}.
\end{aligned} \tag{8.147}$$

Since the γ pole of (8.142) is (from (8.137) + (8.138))

$$3\gamma\epsilon^5(m-n)(m-2n-1)R_m z^{m-2n-k_1-2}$$

we obtain a second (and final contribution) equal to

$$\begin{aligned}&9\gamma\delta\epsilon^9(m-n)^2(m-2n-1)^2(m-2n-k_2-2)R_m^2 z^{-1}\\&=\frac{9}{2}\gamma\delta\epsilon^9(m-n)^2(m-2n-1)^2(n-k_2-1)R_m^2 z^{-1}.\end{aligned}\tag{8.148}$$

Now we look at the second term of the sixth derivative. Consider the last complex component of the product

$$\begin{aligned}&\hat{Z}_{ttz}(0)\cdot\hat{Z}_{ttz}(0)\\&\quad=\epsilon^8(m-n)^2(m-2n-1)^2R_m^2 z^{2m-4n-4}\\&\qquad+4\gamma\epsilon^7(m-n)^2(m-2n-1)(m-n-k_1-1)R_m^2 z^{2m-3n-k_1-4}\\&\qquad+4\delta\epsilon^6(m-n)^2(m-2n-1)(m-n-k_2-1)R_m^2 z^{2m-3n-k_2-4}\\&\qquad+4\gamma\delta\epsilon^5(m-n)^2(m-2n-1)(m-k_1-k_2-1)R_m^2 z^{2m-2n-k_1-k_2-4}\\&\qquad+8\gamma\delta\epsilon^5(m-n)^2(m-n-k_1-1)(m-n-k_2-1)R_m^2 z^{2m-2n-k_1-k_2-4}.\end{aligned}\tag{8.149}$$

Thus the $\gamma\delta$ term of $\hat{Z}_{ttz}(0)\cdot\hat{Z}_{ttz}(0)\phi_t(0)$ is

$$\begin{aligned}&-4\gamma\delta\epsilon^9(m-n)^2(m-2n-1)(m-n-k_1-1)(n-k_2)R_m^2/z\\&\quad-4\gamma\delta\epsilon^9(m-n)^2(m-2n-1)(m-n-k_2-1)(n-k_1)R_m^2/z\\&\quad-\gamma\delta\epsilon^9(m-n)^2(m-2n-1)^2(2n-k_1-k_2)R_m^2/z.\end{aligned}\tag{8.150}$$

Moreover

$$\begin{aligned}&2\hat{Z}_{ttz}(0)\cdot\hat{Z}_{tz}(0)\phi_{tt}(0)\\&\quad=8\gamma\delta\epsilon^9(m-n)^2(m-2n-1)(n-k_1)(n-k_2)R_m^2/z\\&\quad=4\gamma\delta\epsilon^9(m-n)^2(m-2n-1)(n-k_1)(n-k_2)R_m^2/z\\&\qquad+4\gamma\delta\epsilon^9(m-n)^2(m-2n-1)(n-k_2)(n-k_2)R_m^2/z.\end{aligned}\tag{8.151}$$

Thus, one easily sees that ($k_1=n-1$)

$$\begin{aligned}&\hat{Z}_{ttz}(0)\cdot\hat{Z}_{ttz}(0)\phi_t+2\hat{Z}_{ttz}(0)\cdot\hat{Z}_{tz}(0)\phi_{tt}\\&\quad=-5(m-n)^2(m-2n-1)^2(n-k_2+1).\end{aligned}\tag{8.152}$$

Considering the formula for the sixth derivative, we multiply the sum of (8.146) and (8.148) by 20 and (8.152) by 30 to obtain

$$\begin{aligned}&60\gamma\delta\epsilon^9(m-n)^2(m-2n-1)^2(n-k_2+1)R_m^2/z\\&\quad+90\gamma\delta\epsilon^9(m-n)^2(m-2n-1)^2(n-k_2-1)R_m^2/z\\&\quad-150\gamma\delta\epsilon^9(m-n)^2(m-2n-1)^2(n-k_2+1)R_m^2/z\\&=-180\gamma\delta\epsilon^9(m-n)^2(m-2n-1)^2R_m^2/z.\end{aligned}$$

It is not hard to see that, with this choice of generators, the sixth derivative does not depend on the first complex components. Choosing $\delta < 0, \gamma > 0$ we can ensure monotonicity and make the sixth derivative negative.

We now consider the last cases. We start with the case $2m - 4 = 6n - 2$, or $m = 3n + 1$, and begin by assuming $n > 4$. We choose the generator

$$\tau := \epsilon\gamma c/z^{k_1} + c\delta/z^{k_2},$$

$k_1 = (n-1)$, $k_2 = 3$, $k_2 < k_1$. Then $2m - 5 = 6n - 3 = 6(n-1) + 3$.

In this case, since $A_i = \lambda_i A_1 = \lambda_j(1,i)$, $1 \leq j \leq 2m - 2n$ and the real part of

$$\hat{Z}_{ttz}(0)\phi(0) + 2\hat{Z}_{tz}(0)\phi_t(0) + \hat{Z}_z(0)\phi_{tt}(0) \tag{8.153}$$

(cf. (8.132)) has no poles, it follows that the first complex components of (8.153) have no poles. Since $m > 3(n-1) + 2$, the last complex component also has no poles, and (8.153) is holomorphic.

Using our general formula for odd order derivatives, and recalling that (cf. Lemma 2.4, Chap. 2)

$$\hat{Z}_{ttz}(0) \cdot \hat{Z}_z(0) = -\hat{Z}_{tz}(0) \cdot \hat{Z}_{tz}(0),$$

we obtain the following formula for the seventh derivative of Dirichlet's energy, namely

$$\begin{aligned}
\frac{1}{40}\frac{d^7E}{dt^7}(0) &= \mathrm{Re}\int (1+z^2)[\hat{Z}_{tttz}(0) \cdot \hat{Z}_{tttz}(0)\phi(0)]\,d\theta =: (D)\\
&+ 6\,\mathrm{Re}\int (1+z^2)[\hat{Z}_{tttz}(0) \cdot \hat{Z}_{ttz}(0)\phi_t(0)]\,d\theta =: (E)\\
&+ 6\,\mathrm{Re}\int (1+z^2)[\hat{Z}_{tttz}(0) \cdot \hat{Z}_{tz}(0)\phi_{tt}(0)]\,d\theta =: (F)\\
&+ 2\,\mathrm{Re}\int (1+z^2)[\hat{Z}_{tttz}(0) \cdot \hat{Z}_z(0)\phi_{ttt}(0)]\,d\theta =: (G)\\
&+ \left(\frac{180}{40}\right)\mathrm{Re}\int (1+z^2)[\hat{Z}_{ttz}(0) \cdot \hat{Z}_{ttz}(0)\phi_{tt}(0)]\,d\theta =: (H)\\
&+ \left(\frac{240}{40}\right)\mathrm{Re}\int (1+z^2)[\hat{Z}_{ttz}(0) \cdot \hat{Z}_{tz}(0)\phi_{ttt}(0)]\,d\theta =: (I).
\end{aligned} \tag{8.154}$$

We begin our calculations, again calculating only the last complex components:

$$\hat{Z}_{tz}(0) = \epsilon\gamma(m-k_1)R_m z^{m-k_1-1} + \delta(m-k_2)R_m z^{m-k_2-1} \tag{8.155}$$

and

$$\begin{aligned}
-\phi_t(0) &= \epsilon^2\gamma^2(n-k_1)R_m z^{-2(k_1+1)} + \delta^2(n-k_2)R_m z^{-(2k_2+1)}\\
&+ \epsilon\gamma\delta(2n-k_1-k_2)R_m z^{-(k_1+k_2+1)} + \cdots.
\end{aligned} \tag{8.156}$$

We may further ignore the $\delta^2, \delta^3, \ldots$ terms. Thus

$$\hat{Z}_{tz}(0)\phi + \hat{Z}_z(0)\phi_t(0) = \epsilon^2\gamma^2(m-n)R_m z^{m-2k_1-1} + 2\epsilon\gamma\delta(m-n)R_m z^{m-k_1-k_2-1} + \cdots \quad (8.157)$$

whence

$$\hat{Z}_{ttz}(0) = 2\gamma\delta\epsilon(m-n)(m-k_1-k_2-1)R_m z^{m-k_1-k_2-2} + \epsilon^2\gamma^2(m-n)(m-2k_1-1)R_m z^{m-2k_1-2} + \cdots \quad (8.158)$$

and

$$\begin{aligned}\hat{Z}_{ttz}(0)\phi &= \epsilon^3\gamma^3(m-n)(m-2k_1-1)R_m z^{m-3k_1-2} \\ &\quad + 2\gamma^2\delta\epsilon^3(m-n)(m-k_1-k_2-1)R_m z^{m-k_1-k_2-2} \\ &\quad + \epsilon^2\gamma^2\delta(m-n)(m-2k_1-1)R_m z^{m-k_1-k_2-2} + \cdots. \end{aligned} \quad (8.159)$$

We need to evaluate

$$\hat{Z}_{ttz}(0)\phi(0) + 2\hat{Z}_{tz}(0)\phi_t(0) + \hat{Z}_z(0)\phi_{tt}(0).$$

First,

$$\begin{aligned}-2\hat{Z}_{tz}(0)\phi_t(0) &= 2\epsilon^3\gamma^3(m-k_1)(n-k_1)R_m z^{m-3k_1-2} \\ &\quad + 2\epsilon^2\gamma^2\delta(m-k_2)(n-k_1)R_m z^{m-2k_1-k_2-2} \\ &\quad + 2\epsilon^2\gamma^2\gamma(m-k_1)(2n-k_1-k_2)R_m z^{m-2k_1-k_2-2} + \cdots. \end{aligned} \quad (8.160)$$

Noting that (cf. (8.129)) $\phi_{tt}(0) = 2\tau(\tau_x + n\tau/x)^2 + \cdots$ we have

$$\begin{aligned}\phi_{tt}(0) &= 4\gamma^2\delta\epsilon^2(n-k_1)(n-k_2)z^{-(2k_1+k_2+2)} + 2\gamma^2\delta\epsilon^2(n-k_1)^2 z^{-(2k_1+k_2+2)} \\ &\quad + 2\gamma^3\epsilon^3(n-k_1)^2 z^{-3k_1-2} + \cdots. \end{aligned} \quad (8.161)$$

Thus (8.153) equals

$$(\mathrm{A}): \begin{cases} 2\gamma^2\delta\epsilon^2(m-n)(m-k_1-k_2-1)R_m z^{m-2k_1-k_2-2} \\ \quad + \gamma^2\delta\epsilon^2(m-n)(m-2k_1-1)R_m z^{m-2k_1-k_2-2} \\ \quad + \gamma^2\epsilon^3(m-n)(m-2k_1-1)R_m z^{m-3k_1-2} \end{cases} \quad (8.162)$$

$+$

$$(\mathrm{B}): \begin{cases} -2\gamma^2\epsilon^2\delta(m-k_2)(n-k_1)R_m z^{m-2k_1-k_2-2} \\ \quad - 2\gamma^2\epsilon^2\delta(m-k_1)(2n-k_1-k_2)R_m z^{m-2k_1-k_2-2} \\ \quad - 2\gamma^3\epsilon^3(m-k_1)(n-k_1)R_m z^{m-3k_1-2} \end{cases} \quad (8.163)$$

$+$

$$(\mathrm{C}): \begin{cases} 4\gamma^2\epsilon^2\delta(n-k_1)(n-k_2)R_m z^{m-2k_1-k_2-2} \\ \quad + 2\gamma^2\epsilon^2\delta(n-k_1)^2 R_m z^{m-2k_1-k_2-2} \\ \quad + 2\gamma^2\epsilon^2\delta(n-k_1)^2 R_m z^{m-3k_1-2}. \end{cases} \quad (8.164)$$

Adding (A) + (B) + (C) we see that (8.153) equals

$$\begin{aligned}&3\gamma^2\epsilon^2\delta(m-n)(m-2n-1)R_m z^{m-2k_1-k_2-2}\\&\quad+\gamma^3\epsilon^3(m-n)(m-2n-1)R_m z^{m-3k_1-2}\end{aligned}\tag{8.165}$$

yielding that

$$\begin{aligned}\hat{Z}_{tttz}(0)&=3\gamma^2\epsilon^2\delta(m-n)(m-2n-1)(m-2k_1-k_2-2)R_m z^{m-2k_1-k_2-3}\\&\quad+\gamma^3\epsilon^3(m-n)(m-2n-1)(m-3k_1-2)R_m z^{m-3k_1-3}+\cdots.\end{aligned}\tag{8.166}$$

Thus, again ignoring $\delta^2, \delta^3, \ldots$ terms

$$\begin{aligned}&\hat{Z}_{tttz}(0)\cdot\hat{Z}_{tttz}(0)\\&\quad=6\gamma^5\epsilon^5\delta(m-n)^2(m-2n-1)^2(m-2k_1-k_2-2)(m-3k_1-2)\\&\qquad\cdot R_m z^{2m-5k_1-k_2-6}\\&\qquad+\gamma^6\epsilon^6(m-n)^2(m-2n-1)^2(m-3k_1-2)^2R_m z^{2m-6k_1-6}.\end{aligned}\tag{8.167}$$

This implies that the integrand of (D) is

$$\begin{aligned}&\hat{Z}_{tttz}(0)\cdot\hat{Z}_{tttz}(0)\phi(0)\\&\quad=6\gamma^6\epsilon^6\delta(m-n)^2(m-2n-1)^2(m-2k_1-2)(m-3k_1-2)R_m z^{2m-6k_1-k_2-6}\\&\qquad+\gamma^6\epsilon^6\delta(m-n)^2(m-2n-1)^2(m-3k_1-2)^2R_m z^{2m-6k_1-k_2-6}\\&\qquad+\gamma^7\epsilon^7(m-n)^2(m-2n-1)^2(m-3k_1-2)^2R_m z^{2m-7k_1-6}+\cdots.\end{aligned}\tag{8.168}$$

Using (8.166) and (8.159) we see that

$$\begin{aligned}&\hat{Z}_{tttz}(0)\cdot\hat{Z}_{ttz}(0)\\&\quad=2\gamma^4\epsilon^4\delta(m-n)^2(m-2n-1)(m-k_1-k_2-1)(m-3k_1-2)R_m z^{2m-4k_1-k_2-5}\\&\qquad+3\gamma^4\epsilon^4\delta(m-n)^2(m-2n-1)(m-2k_1-k_2)(m-2k_1-1)R_m z^{2m-4k_1-k_2-5}\\&\qquad+\gamma^5\epsilon^5(m-n)^2(m-2n-1)(m-3k_1-2)(m-2k_1-1)R_m z^{2m-5k_1-5}+\cdots.\end{aligned}\tag{8.169}$$

Thus, we obtain the integrand of (E):

$$\begin{aligned}&6\hat{Z}_{tttz}(0)\cdot\hat{Z}_{ttz}(0)\phi_t(0)\\&\quad=-(\gamma^6\epsilon^6\delta)\cdot\{12(m-n)^2(m-2n-1)(m-k_1-k_2-1)\\&\qquad\cdot(m-3-k_1-2)(n-k_1)\\&\qquad+18(m-n)^2(m-2n-1)(m-2k_1-k_2-2)(m-2k_1-1)(n-k_1)\\&\qquad+6(m-n)^2(m-2n-1)(m-3k_1-2)(m-2k_1-1)(2n-k_1-k_2)\}\\&\qquad\cdot R_m z^{2m-6k_1-k_2-6}\\&\qquad-6\gamma^6\epsilon^6(m-n)^2(m-2n-1)(m-3k_1-2)(m-2k_1-1)(n-k_1)\\&\qquad\cdot R_m z^{2m-7k_1-6}+\cdots.\end{aligned}\tag{8.170}$$

Furthermore, working towards the integrand of (F) we have

$$
\begin{aligned}
&6\hat{Z}_{tttz}(0)\cdot\hat{Z}_{tz}(0)\\
&\quad=\gamma^3\epsilon^3\delta\{18(m-n)(m-2n-1)(m-2k_1-k_2-2)(m-k_1)\\
&\qquad+6(m-n)(m-2n-1)(m-3k_1-2)(m-k_2)\}R_m z^{2m-3k_1-k_2-4}\\
&\qquad+6\gamma^4\epsilon^4(m-n)(m-2n-1)(m-3k_1-2)(m-k_1)R_m z^{2m-4k_1-4}.
\end{aligned}
\tag{8.171}
$$

Using formula (8.161) for $\phi_{tt}(0)$ we get the integrand of (F):

$$
\begin{aligned}
&\hat{Z}_{tttz}(0)\cdot\hat{Z}_{tz}(0)\phi_{tt}(0)\\
&\quad=\gamma^6\epsilon^6\delta\{36(m-n)(m-2n-1)(m-2k_1-k_2-2)(m-k_1)(n-k_1)^2\\
&\qquad+12(m-n)(m-2n-1)(m-3k_1-2)(m-k_2)(n-k_1)^2\\
&\qquad+24(m-n)(m-2n-1)(m-3k_1-2)(m-k_1)(n-k_1)(n-k_2)\\
&\qquad+12(m-n)(m-2n-1)(m-3k_1-2)(m-k_1)(n-k_1)^2\}R_m z^{2m-6k_1-k_2-6}\\
&\qquad+12\gamma^7\epsilon^7(m-n)(m-2n-1)(m-3k_1-2)(m-k_1)(n-k_1)^2R_m z^{2m-7k_1-6}.
\end{aligned}
\tag{8.172}
$$

We now evaluate the integrand of (G). As, in the case of $\phi_t(0)$ and $\phi_{tt}(0)$ it is not hard to see that

$$
\phi_{ttt}(0)=6\tau(\tau_x+n\tau/x)^3+\cdots \tag{8.173}
$$

yielding that

$$
\begin{aligned}
\phi_{ttt}(0)&=-18\gamma^3\epsilon^3\delta(n-k_1)^2(n-k_2)z^{-3k_1-k_2-3}\\
&\quad-6\gamma^3\epsilon^3\delta(n-k_1)^3z^{-3k_1-k_2-3}-6\gamma^4\epsilon^4(n-k_1)^3z^{-4k_1-3}.
\end{aligned}
\tag{8.174}
$$

Hence, we obtain the integrand of (G) as

$$
\begin{aligned}
&-\gamma^6\epsilon^6\delta\{36(m-n)(m-2n-1)(m-2k_1-k_2-2)(n-k_1)^3\\
&\quad+36(m-n)(m-2n-1)(m-3k_1-2)(n-k_1)^2(n-k_2)\\
&\quad+12(m-n)(m-2n-1)(m-3k_1-2)(n-k_1)^3\}R_m z^{2m-6k_1-k_2-6}\\
&\quad-12\gamma^7\epsilon^7(m-n)(m-2n-1)(m-3k_1-2)(n-k_1)^3R_m z^{2m-7k_1-6}.
\end{aligned}
\tag{8.175}
$$

Adding the integrands of first (G) to (F) and then this sum to (E) and then to (D) we get the sum (G) + (F) + (E) + (D) equal to

$$
\begin{aligned}
&\gamma^6\epsilon^6\delta\{-18(m-n)^2(m-2n-1)^2(m-3k_1-2)(n-k_1)\\
&\quad-18(m-n)^2(m-2n-1)^2(m-2k_1-k_2-2)(n-k_1)\\
&\quad-6(m-n)^2(m-3k_1-2)(m-2k_2-1)(n-k_2)\\
&\quad+6(m-n)^2(m-2n-1)^2(m-2k_1-k_2-2)(m-3k_1-2)
\end{aligned}
$$

$$
\begin{aligned}
&+ (m-n)^2(m-2n-1)^2(m-3k_1-2)^2\}R_m z^{2m-6k_1-k_2-6} \\
&+ \gamma^7\epsilon^7\{(m-n)^2(m-2n-1)^2(m-3k_1-2)^2 \\
&- 6(m-n)^2(m-2n-1)(m-3k_1-2)(m-2k_1-1)(n-k_1) \\
&+ 12(m-n)(m-2n-1)(m-3k_1-2)(m-k_1)(n-k_1)^2 \\
&- 12(m-n)(m-2n-1)(m-3k_1-2)(n-k_1)^3\}R_m z^{2m-7k_1-6}. \qquad (8.176)
\end{aligned}
$$

Noting that $n-k_1=1, n-k_2=n-3, m=3n+1, m-3k_1-2=2, m-2k_1-1=n+2, m-2k_1-k_2-2=n-2$, we see that the integrand of (E) + (F) + (G) + (D) equals

$$
\begin{aligned}
&\gamma^6\epsilon^6\delta\{-36(m-n)^2(m-2n-1)^2 - 18(m-n)^2(m-2n-1)^2(n-2) \\
&\quad - 12(m-n)^2(n+2)(n-3) + 12(m-n)^2(m-2n-1)^2(n+2) \\
&\quad + 4(m-n)^2(m-2n-1)^2\}R_m z^{2m-6k_1-k_2-6} \qquad (8.177) \\
&\quad + \gamma^7\epsilon^7\{4(m-n)^2(m-2n-1)^2 - 12(m-n)^2(m-2n-1)(n+2) \\
&\quad + 48(m-n)(m-2n-1)(n+1) \\
&\quad - 24(m-n)(m-2n-1)\}R_m z^{2m\ \ 7k_1-6} + \cdots. \qquad (8.178)
\end{aligned}
$$

Let us now focus on the coefficients of $\gamma^6\epsilon^6\delta R_m z^{2m-6k_1-k_2-6}$. Rewriting these coefficients we have

$$
\begin{aligned}
&-(m-n)^2(m-2n-1)^2(36) \\
&\quad - (m-n)^2(m-2n-1)^2(18n-36) \\
&\quad - 12(m-n)^2(n+2)(n-3) \\
&\quad + (m-n)^2(m-2n-1)^2(12n+28) \qquad (8.179)
\end{aligned}
$$

which equals

$$
\begin{aligned}
&-(m-n)^2(m-2n-1)^2(6n-28) \\
&\quad - 12(m-n)^2(n+2)(n-3), \qquad (8.180)
\end{aligned}
$$

which is the integrand of (E) + (F) + (G) + (D).

Calculating the coefficient of $\gamma^6\epsilon^6\delta R_m z^{2m-6k_1-k_2-6}$ in $\hat{Z}_{ttz}(0)\cdot\hat{Z}_{ttz}(0)\phi_{tt}(0)$, the integrand of (H), we obtain

$$
\begin{aligned}
&4(m-n)^2(m-2k_1-1)^2(n-k_1)(n-k_2) \\
&\quad + 2(m-n)^2(m-2k_1-1)^2(n-k_1)^2 \\
&\quad + 8(m-n)^2(m-2k_1-1)(m-k_1-k_2-1)(n-k_1) \qquad (8.181)
\end{aligned}
$$

which simplifies to

$$
\begin{aligned}
&4(m-n)^2(n+2)^2(n-3) \\
&\quad + 2(m-n)^2(n+2)^2 \\
&\quad + 16(m-n)^2(n+2)(n-1). \qquad (8.182)
\end{aligned}
$$

Multiplying this by (180/40) we obtain

$$\begin{aligned}
&18(m-n)^2(n+2)^2(n-3)^2 \\
&\quad + 9(m-n)^2(n+2)^2 \\
&\quad + 72(m-n)^2(n+2)(n-1).
\end{aligned} \tag{8.183}$$

Now the coefficient of $\gamma^6\epsilon^6\delta R_m^2 z^{2m-6k_1-k_2-6}$ in $\hat{Z}_{ttz}(0)\cdot\hat{Z}_{tz}(0)\phi_{ttt}(0)$, the integrand of (I) is

$$\begin{aligned}
&-6(m-n)(m-2k_1-1)(m-k_2)(n-k_1)^3 \\
&\quad - 36(m-n)(m-k_1)(m-k_1-k_2-1)(n-k_1)^2(n-k_2) \\
&\quad - 18(m-n)(m-2k_1-1)(m-k_1)(n-k_1)^2(n-k_2) \\
&\quad - 12(m-n)(m-k_1)(m-k_1-k_2-1)(n-k_1)^3 \\
&\quad - 6(m-n)(m-2k_1-1)(m-k_2)(n-k_1)^3
\end{aligned} \tag{8.184}$$

which simplifies to

$$-(m-n)(180n^3-366n^2-372n+144). \tag{8.185}$$

Multiplying this by 6 we get

$$-(m-n)(1080n^3-2196n^2-2232n+1464). \tag{8.186}$$

Simplifying (8.183) we have

$$(m-n)(36n^4+90n^3-684n-324). \tag{8.187}$$

Simplifying (8.180) we have

$$-(m-n)(12n^4-72n^3-40n^2-158n-72). \tag{8.188}$$

Summing (8.186)–(8.188) we obtain

$$(m-n)(24n^4-936n^3+2200n^2+1706n-1706). \tag{8.189}$$

We wish to show that this cannot vanish for even n. So set $n=2p$, then (8.189) equals

$$4(m-n)(384p^4-7484p+8800p^2+3412p-1706). \tag{8.190}$$

With our chosen generators, there is a contribution from the first complex components arising from the product $A_1\cdot A_{2m-2n+1}$ in the terms $Z_{tttz}\cdot Z_{tz}\phi_{tt}$, $Z_{tttz}\cdot Z_z\phi_{ttt}$, and $Z_{ttz}\cdot Z_{tz}\phi_{ttt}$. No first complex components contributions arise from terms containing t-derivatives of order higher than one. We note that the contribution of these terms is divisible by four. A straightforward calculation shows that the sum of this contribution and (8.190) is

$$24n^4-936n^3+7564n^2+3482n-13322. \tag{8.191}$$

We need to show that for n even this cannot be zero. Since n is even, every term in (8.191), except the last, is divisible by four, and so (8.191) can never vanish.

If (8.191) should be negative, we choose $\gamma > 0$ and $\gamma < 0$; if positive, $\gamma > 0$ and $\delta > 0$. In either case, we have a monotonic variation which makes the seventh derivative negative for $2m - 4 = 6n - 2$, $n > 4$.

We thus have three remaining cases, $n = 2$, $m = 4$, $m = 7$ and $n = 4$, $m = 13$. In the first, with generator $\tau = \gamma c/x$, the fourth and all lower order derivatives vanish, and the fifth derivative is negative. For $n = 2$, $m = 7$ with the same generator the eighth derivative can be made negative. For $n = 4$, $m = 13$ with the generator $\tau = \gamma c/x^3$, the seventh derivative can be made negative.

Thus, such branched minimal surfaces cannot be minima. This concludes the proof of the fact that C^∞ minimal surfaces which have a non-exceptional branch point spanning a contour with non-zero torsion and curvature, cannot be weak relative minima for Dirichlet's energy or area in any C^r topology.

The culmination of the work of the preceding chapters shows that in order to prove that minimal surfaces with an exceptional branch point cannot be a minima requires the existence of a Taylor expansion with the property that, after a change of variables, $k > l$ and there exists an $\overline{m} > m$, with $R_{\overline{m}} \not\equiv 0$ and $1 + \overline{m}$ not an integer multiple of $1 + n$ without a convergent Taylor series (the analytic case); this appears to be impossible.

In fact, if all $R_j = 0$, $j > m$, or if for all $j > m$, the only non-zero R_j are those for which $j + 1 = k(n + 1)$, $k \in N$, then, using the generator $1/z^n$, the methods of this book would yield a curve along which all of the derivatives of Dirichlet's energy are zero (one chooses $\phi_t = 1/z^n$ for all t small; thus $D_t^\beta \phi(0) = 0$ for all $\beta > 0$, and the result easily follows). Thus, in the exceptional case, finding criteria to determine whether or not the surface is a minimum is analogous to finding criteria to determine if 0 is a minimum for

$$f(x) = \begin{cases} e^{-1/x^2}, & x > 0, \\ 0, & x = 0, \\ \pm e^{-1/x^2}, & x < 0. \end{cases}$$

In this regard we refer the reader to the example of Gulliver [4], a minimal surface with an atypical boundary branch point. It is unknown if this surface is a minimum.

In this book we have put forward the argument that forced Jacobi fields are the key to understanding the fact that branched minimal surfaces cannot, in most cases, be minima. If this insight is correct, then from the above remarks, it would appear that for a C^∞, but not analytic, minimal surface $\hat{X}$ with an exceptional boundary branch point, the question of whether or not $\hat{X}$ can be a minimum is not decidable. So it is perhaps only in this situation that the original guesses by Douglas and Courant (see Scholia) concerning the existence of branch points for minimizers may turn out to be correct.

The analytic case, on the other hand, was, as previously mentioned, worked out by White [1], who showed that analytic surfaces in $C(\Gamma)$ with a true boundary branch point cannot be minima. The methods of this book apply to this case. First, if $\hat{X} \in C(\Gamma)$, $\hat{X}$, by Tomi's proof (Sect. 6.1) cannot have an analytically false boundary branch point. Thus, the integer Γ (Chap. 5) is defined and one can use interior methods to prove the absence of branch points for C^0 weak relative minima in the exceptional case also.

Chapter 9
Scholia

The solution of Plateau's problem presented by J. Douglas [1] and T. Radó [1] was achieved by a – very natural – redefinition of the *notion of a minimal surface* $X : \Omega \to \mathbb{R}^3$ which is also used in our book[1]: Such a surface is a harmonic and conformally parametrized mapping; but it is not assumed to be an immersion. Consequently X may possess branch points, and thus some authors speak of "branched immersions". This raises the question whether or not Plateau's problem always has a solution which is immersed, i.e. regular in the sense of differential geometry. Certainly there exist minimal surfaces with branch points; but one might conjecture that area minimizing solutions of Plateau's problem are free of (interior) branch points. To be specific, let Γ be a closed, rectifiable Jordan curve in $\mathbb{R}^3$, and denote by $\mathcal{C}(\Gamma)$ the class of disk-type surfaces $X : B \to \mathbb{R}^3$ bounded by Γ which was defined in Chap. 1. Then one may ask: *Suppose that $X \in \mathcal{C}(\Gamma)$ is a disk-type minimal surface $X : \overline{B} \to \mathbb{R}^3$ which minimizes both A and D in $\mathcal{C}(\Gamma)$. Does X have branch points in B (or in $\overline{B}$)?*

Radó [1], pp. 791–795, gave a first answer to this question for some special classes of boundary contours Γ, using the following result:

If $X_w(w)$ vanishes at some point $w_0 \in B$ then any plane through the point $P_0 := X(w_0)$ intersects Γ in at least four distinct points.

This observation has the following interesting consequence: *Suppose that there is a straight line $\mathcal{L}$ in $\mathbb{R}^3$ such that any plane through $\mathcal{L}$ intersects Γ in at most two distinct points. Then any minimal surface $X \in \mathcal{C}(\Gamma)$ has no branch points in B.* In fact, for $P_0 \notin \mathcal{L}$, the plane Π determined by P_0 and $\mathcal{L}$ meets Γ in at most two points, and for $P_0 \in \mathcal{L}$ there are infinitely many such planes.

In particular: *If Γ has a simply covered star-shaped image under a (central or parallel) projection upon some plane Π_0, then any minimal surface $X \in \mathcal{C}(\Gamma)$ is free of branch points in B.*

Somewhat later, Douglas [2], pp. 733, 739, 753, thought that he had found a contour Γ with the property that any minimal surface $X \in \mathcal{C}(\Gamma)$ is branched, namely

[1] We now denote a minimal surface by X and no longer by $\hat{X}$, i.e. we no longer emphasize the difference between a surface $\hat{X}$ and its boundary values X.

A. Tromba, *A Theory of Branched Minimal Surfaces*,
Springer Monographs in Mathematics,
DOI 10.1007/978-3-642-25620-2_9, © Springer-Verlag Berlin Heidelberg 2012

a curve whose orthogonal projection onto the x^1, x^2-plane is a certain closed curve with a double point.

Radó [2], p. 109, commented on this assertion as follows: A curve Γ with this x^1, x^2-projection can be chosen in such a way that its x^1, x^3-projection is a simply covered star-shaped curve in the x^1, x^3-plane; thus no minimal surface in $\mathcal{C}(\Gamma)$ has a branch point.

In 1941, Courant [1] is believed to have found a contour Γ for which some minimizer of Dirichlet's integral in $\mathcal{C}(\Gamma)$ has an interior branch point. This assertion is not correct, as Osserman [2], p. 567, pointed out in 1970. Moreover, in [2] he described an ingenious line of argument which seemed to exclude interior branch points for area minimizing solutions of Plateau's problem. For this purpose he distinguished between *true* and *false* branch points (cf. Osserman [1], p. 154, Definition 6; and, more vaguely, [2], p. 558): A branch point is false, if the image of some neighbourhood of the branch point lies on a regularly embedded minimal surface; otherwise it is a true branch point. Osserman's treatment of the false branch points is incomplete, but contains essential ideas used by later authors, while his exclusion of true branch points is essentially complete (see also Gulliver–Osserman–Royden [1], p. 751, D. Wienholtz [1], p. 2). The principal ideas of Osserman in dealing with true branch points w_0 are the following: First, the geometric behaviour of the minimal surface X in the neighbourhood of w_0 is studied, yielding the existence of branch lines. Then a remarkable discontinuous parameter transformation G is introduced such that $\tilde{X} := X \circ G$ lies again in $\mathcal{C}(\Gamma)$ and has the same area as X, but in addition $\tilde{X}$ has a wedge, and so its area can be reduced by "smoothing out" the wedge. Osserman's definition of G is somewhat sloppy, but K. Steffen has kindly pointed out to us how this can be remedied and the construction of the area reducing surface can rigorously be carried out.

Osserman's paper [2] was the decisive breakthrough in excluding true branch points for area minimizing minimal surfaces in $\mathbb{R}^3$, and it inspired the succeeding papers by R. Gulliver [1] and H.W. Alt [1], [2], which even tackled the more difficult branch point problem for H-surfaces and for minimal surfaces in a Riemannian manifold (Gulliver). Nearly simultaneously, both authors published proofs of the assertion that area minimizing minimal surfaces in $\mathcal{C}(\Gamma)$ possess no interior branch points (and of the analogous statement for H-surfaces).

Gulliver's reasoning runs as follows: Let us assume that $w_0 = 0$ is an interior branch point of the minimal surface $X \in \overline{\mathcal{C}}(\Gamma)$, $X : \overline{B} \to \mathbb{R}^3$. Then there is a neighbourhood $V \subset\subset B$ of 0 in which two oriented Jordan arcs $\gamma_1, \gamma_2 \in C^1([0,1], B)$ exist with $\gamma_1(0) = \gamma_2(0) = 0$, $|\gamma_j'(0)| = 1$, $\gamma_1'(0) \neq \gamma_2'(0)$, $X(\gamma_1(t)) \equiv X(\gamma_2(t))$, and such that $(X_u \wedge X_v)(\gamma_1(t))$, $(X_u \wedge X_v)(\gamma_2(t))$ are linearily independent for $0 < t \leq 1$. One can assume that ∂V is smooth, and that γ_1, γ_2 meet ∂V transversally at distinct points $\gamma_1(\epsilon), \gamma_2(\epsilon)$, $0 < \epsilon < 1$. Then there is a homeomorphism $F : \overline{B}_\epsilon \to \overline{V}$ with $F(it) = \gamma_1(t)$, $F(-it) = \gamma_2(t)$ for $0 \leq t \leq \epsilon$, and $F \in C^2(\overline{B}_\epsilon \setminus \{0\})$ where $B_\epsilon := B_\epsilon(0) = \{w \in \mathbb{C} : |w| < \epsilon\}$. Define a discontinuous map $G : \overline{B}_\epsilon \to \overline{B}_\epsilon$ such that $\{it : 0 < t \leq 1\}$ and $\{-it : 0 < t \leq 1\}$ are mapped to i and $-i$ respectively; $\pm\epsilon/2$ are taken to zero; on the segments of discontinuity $[-\epsilon/2, 0]$ and $[0, \epsilon/2]$ are each given two linear mappings by limiting values under approach from the two sides; G is continuous on a neighbourhood of ∂B_ϵ with $G|_{\partial B_\epsilon} = \mathrm{id}_{\partial B_\epsilon}$; and G is conformal on each component of $B_\epsilon \setminus I_\epsilon$\imaginary axis, where I_ϵ is the interval $[-\epsilon/2, \epsilon/2]$ on the real axis. Thus $X \circ F \circ G$ is continuous and piecewise C^2. Now define

$$\overline{X}(w) := \begin{cases} (X \circ F \circ G \circ F^{-1})(w) & \text{for } w \in V, \\ X(w) & \text{for } w \in \overline{B} \setminus V. \end{cases}$$

Then $\overline{X}$ is continuous and piecewise C^2, and $\overline{X} \in \mathcal{C}(\Gamma)$. The metric

$$ds^2 := \langle d\overline{X}, d\overline{X} \rangle = a\, du^2 + 2b\, du\, dv + c\, dv^2,$$
$$a := |\overline{X}_u|^2, \quad b := \langle \overline{X}_u, \overline{X}_v \rangle, \quad c := |\overline{X}_v|^2,$$

induced on B by pulling back the metric induced from $\mathbb{R}^3$ along $\overline{X}$ has bounded, piecewise smooth coefficients. It follows from the uniformization theorem of Morrey ([1], Theorem 3) that there exists $T : B \to B$ with L^2 second derivatives, which is almost everywhere conformal from B with its usual metric to B with its induced metric, and T may be extended to a homeomorphism $\overline{B} \to \overline{B}$.

Now define $\tilde{X} := \overline{X} \circ T$; then $\tilde{X} \in \mathcal{C}(\Gamma)$, $A(\tilde{X}) = A(X)$, and $\langle \tilde{X}_w, \tilde{X}_w \rangle = 0$ a.e. on B, and consequently

$$\inf_{\mathcal{C}(\Gamma)} D = \inf_{\mathcal{C}(\Gamma)} A = D(X) = A(X) = A(\tilde{X}) = D(\tilde{X}).$$

Thus $\tilde{X}$ is D-minimizing, and so its surface normal $\tilde{N}$ is continuous on B. On the other hand, the sets $\overline{X}(B)$ and $\tilde{X}(B)$ are the same, and so $\tilde{X}(B)$ has an edge, whence $\tilde{N}$ cannot be continuous, a contradiction.

This reasoning requires two comments. First, D. Wienholtz in his Diploma thesis [1], p. 3 (published as [2]), noted that Gulliver's discontinuous map $G : \overline{B}_\epsilon \to \overline{B}_\epsilon$ does not exist, since its existence contradicts Schwarz's reflection principle. A remedy of this deficiency would be to set up another definition of G or T, such as used in Alt [1], pp. 360–361, or in Steffen–Wente [1], p. 218, or by a modification of the definition of G in Gulliver–Lesley [1], p. 24.

Secondly, the application of one of Morrey's uniformization theorems from [1] is not justified, as this requires besides $a, b, c \in L^\infty(B)$ the assumption

$$ac - b^2 = 1, \tag{*}$$

and this demands the existence of constants $\lambda_1, \lambda_2 \in \mathbb{R}$ with $0 < \lambda_1 \le \lambda_2$ such that

$$\lambda_1[\xi^2 + \eta^2] \le a(w)\xi^2 + 2b(w)\xi\eta + c(w)\eta^2 \le \lambda_2[\xi^2 + \eta^2] \tag{**}$$

for all $(\xi, \eta) \in \mathbb{R}^2$ and for almost all $w \in B$. However, $\overline{X}(w) \equiv X(w)$ on $B \setminus V$, and X might have another branch point $w_0' \in B \setminus V$; then $a(w_0') = b(w_0') = c(w_0') = 0$, and so neither (*) nor (**) were satisfied.

There is another possibility to correct this deficiency. Suppose that $\overline{X}$ is quasiconformal in the sense that

$$|\overline{X}_u|^2 + |\overline{X}_v|^2 \le \kappa |\overline{X}_u \wedge \overline{X}_v| \quad \text{(a.e. on } B)$$

holds for some constant $\kappa > 0$. Then it follows

$$a, |b|, c \le \kappa \sqrt{ac - b^2},$$

and thus the quadratic form

$$d\sigma^2 := \alpha\, du^2 + 2\beta\, du\, dv + \gamma\, dv^2$$

with

$$\alpha := \frac{a}{\sqrt{ac - b^2}}, \qquad \beta := \frac{b}{\sqrt{ac - b^2}}, \qquad \gamma := \frac{c}{\sqrt{ac - b^2}}$$

satisfies $|\alpha|, |\beta|, |\gamma| \le \kappa$ and $\alpha\gamma - \beta^2 = 1$. Hence one can apply Morrey's first uniformization theorem (as quoted above), obtaining a homeomorphism T from $\overline{B}$ onto $\overline{B}$ with $T, T^{-1} \in H_2^1(B, B)$ such that the pull-back $T^* d\sigma^2$ is a multiple of the Euclidean metric ds_e^2, i.e.

$$T^* d\sigma^2 = \lambda\, ds_e^2$$

whence

$$T^* ds^2 = \tilde{\lambda}\, ds_e^2$$

with $\tilde{\lambda} := \lambda\sqrt{\tilde{a}\tilde{c} - \tilde{b}^2}$, $\tilde{a} := a \circ T$, $\tilde{b} := b \circ T$, $\tilde{c} := c \circ T$.

Now one can proceed for $\tilde{X} := \overline{X} \circ T$ as above. It remains a question as to whether $\overline{X}$ can be constructed in such a way that it is quasiconformal. This would be the case for Gulliver's construction described before, except that Gulliver's G cannot exist on account of the reflection principle. Hence the definition of G must be modified, say, as in Gulliver–Lesley [1]. Then, one might proceed as follows: One decomposes $B_\epsilon \setminus [-\epsilon/2, \epsilon/2]$ in finitely many triangular domains E such that the mappings $G|_{\overline{E}}$ are C^1-diffeomorphisms. Choosing F appropriately, one has to convince oneself that $X \circ \Phi$ with $\Phi := F \circ G \circ F^{-1}$ is quasiconformal if X is a.e. conformal and $|D\Phi|, |D\Phi^{-1}| \le$ const.

Alt's method to exclude true branch points (worked out in detail by D. Wienholtz [1], [2]) eventually uses the same contradiction argument as Gulliver, namely to derive the existence of an energy minimizer $\tilde{X} \in \mathcal{C}(\Gamma)$ with a discontinuous normal $\tilde{N}$. The construction of $\tilde{X}$ is different from Gulliver's approach. Alt defines a new surface $\overline{X}$ on B_ϵ which is quasiconformal, and by reparametrization a new surface $\tilde{X} = \overline{X} \circ \tau$ is obtained which is energy minimizing with respect to its boundary values. Here Morrey's lemma on ϵ-conformal mappings is used as well as an elaboration of Lemma 9.3.3 in Morrey [1].

The non-existence of false branch points for solutions X of Plateau's problem was proved by R. Gulliver [2], H.W. Alt [2], and then by Gulliver–Osserman–Royden in their fundamental 1973 paper [1]. Here one only needs that $X|_{\partial B}$ is one to one, and this observation is used by Alt as well as by Gulliver–Osserman–Royden, while Gulliver also employs the minimizing property of X. K. Steffen pointed out to us that Osserman's original paper [2] already contains significant contributions to the problem of excluding false branch points, and it even is satisfactory if, for some reason, an inner point of X cannot lie on the boundary curve Γ, say, if Γ lies on the surface of a convex body. It should be mentioned that Gulliver [1] in the proof of this Theorem 5.1, Case I ($S = \emptyset$), once again uses Morrey's uniformization theorem.

Furthermore, in Sect. 6 of their paper, Gulliver–Osserman–Royden proved a rather general result on branched surfaces $X : \overline{B} \to \mathbb{R}^n$, $n \ge 2$, such that $X|_{\partial B}$ is injective, which implies the following: *A minimal surface $X \in \mathcal{C}(\Gamma)$ has no false boundary branch points* (see [1], pp. 799–809, in particular Theorem 6.16).

In 1973, R. Gulliver and F.D. Lesley [1] published the following result which we cite in a slightly weaker form: *If Γ is a real analytic and regular contour in $\mathbb{R}^3$, then any area minimizing minimal surface in $\mathcal{C}(\Gamma)$ has no boundary branch points.*

To prove this result they extend a minimizer X across the boundary of the parameter domain B as a minimal surface, so that a branch point w_0 on ∂B can be treated as an inner point. Then the same analysis of X in a small neighbourhood of w_0 can be carried out, and w_0 is either seen to be false or true. To exclude the possibility of a true branch point, they apply the method from Gulliver's paper [1], except that a new discontinuous "Osserman-type" mapping G is described, which is appropriate for this situation. A detailed presentation of this approach or, what might be easier, of Osserman's reasoning applied to boundary branch points, would be desirable. In a different way, the latter was worked out by B. White [1], see below.

The elimination of the possibility of false branch points in the Gulliver–Lesley paper is achieved by using results from the theory of *"branched immersions"*, created by Gulliver, Osserman, and Royden. In fact, one can even apply the result on false boundary branch points quoted above.

The theory of branched immersions was extended by Gulliver [2], [3], [5] in such a way that it applies to surfaces of higher topological type (minimal surfaces and H-surfaces in a Riemannian manifold).

K. Steffen and H. Wente [1] showed in 1978 that minimizers of

$$E_Q(X) := \int_B \left[\frac{1}{2}|\nabla X|^2 + Q(X)\cdot(X_u \wedge X_v)\right] du\,dv$$

in $\mathcal{C}(\Gamma)$ subject to a volume constraint $V(X) = \text{const}$ with

$$V(X) := \frac{1}{3}\int_B X\cdot(X_u \wedge X_v)\,du\,dv$$

have no interior branch points. Their work in particular applies to minimal surfaces. While their treatment of true branch points essentially follows Osserman [2], they simplified, in their special situation, the discussion of false branch points by Gulliver–Osserman–Royden [1] and Gulliver [2].

In 1980, Beeson [1] showed that a minimal surface in $\mathcal{C}(\Gamma)$, given by a local Weierstrass representation, cannot have a true interior branch point if it is a C^1-local minimizer of D in $\mathcal{C}(\Gamma)$. (According to D. Wienholtz, Beeson's proof does not work for C^k-local minimizers with $k \ge 2$.) In this paper Beeson considers higher order derivatives of a localized energy. Later on, in 1994, M. Micallef and B. White [1] excluded the existence of true interior branch points for area minimizing minimal surfaces in a Riemannian 3-manifold, and in 1997, B. White [1] proved that an area minimizing minimal surface $X : \overline{B} \to \mathbb{R}^n, n \ge 3$, cannot have a true branch point on any part of ∂B which is mapped by X onto a real analytic portion of Γ, even if $n \ge 4$. This is quite surprising as X may have interior branch points if $n \ge 4$ (Federer's examples). However, White pointed out that, for any $k < \infty$, one can find C^k-curves Γ in $\mathbb{R}^4$ that bound area minimizing disk-type minimal surfaces with true boundary branch points, and Gulliver [4] found a C^∞-curve in $\mathbb{R}^6$ bounding an area minimizer with a true boundary branch point.

It is a major open question to decide whether or not an area minimizing minimal surface of disk-type in $\mathbb{R}^3$ can have a boundary branch point assuming that it is bounded by a (regular) C^k- or C^∞-contour Γ, rather than by an analytic one.

We furthermore mention the paper of H.W. Alt and F. Tomi [1] where the non-existence of branch points for minimizers to certain free boundary problems is proved and the work of R. Gulliver and F. Tomi [1] where the absence of interior branch points for minimizers of higher genus is established. Specifically, they showed that such a minimizer $X : M \to N$ cannot possess false branch points if X induces an isomorphism on fundamental groups.

In 1977–81, R. Böhme and A. Tromba [1], [2] showed that, *generically*, every smooth Jordan curve in $\mathbb{R}^n$, $n \geq 4$, bounds only immersed minimal surfaces, and admits only simple interior branch points for $n = 3$, but no boundary branch points. "Generic" means that there is an open and dense subset in the space of all sufficiently smooth $\alpha : S^1 \to \mathbb{R}^n$ defining a Jordan curve Γ, for which subset the assertion holds. This result is based on the Böhme–Tromba index theory, which is presented in Dierkes, Hildebrandt and Tromba [2].

A completely new method to exclude the existence of branch points for *minimal surfaces in $\mathbb{R}^3$ which are weak relative minimizers of D* was developed by A.J. Tromba [1] in 1993 by deriving an *intrinsic third derivative of D in the direction of forced Jacobi fields*. He showed that if $X \in \mathcal{C}(\Gamma)$ has only simple interior branch points satisfying a *Schüffler condition* (a condition which had been identified as generic by K. Schüffler [1]), then the third variation of D can be made negative, while the first and second derivatives are zero, and so X cannot be a weak relative minimizer of D in $\mathcal{C}(\Gamma)$. D. Wienholtz in his Doctoral thesis [3] generalized Tromba's method to interior and boundary branch points of arbitrary order, satisfying a "Schüffler-type condition", by computing the third derivative of D in suitable directions generated by forced Jacobi fields. We note that Wienholtz's results also refer to boundary branch points of minimal surfaces in $\mathbb{R}^n, n \geq 3$, but they do not apply to Gulliver's $\mathbb{R}^6$-example (see Wienholtz [3], p. 244).

These results raised the question whether branch points of any order could possibly be excluded by looking at even higher order derivatives of Dirichlet's integral. Such an approach was quite new since variations of higher order for multiple integrals had rarely been studied. In fact, S.S. Chern was known to have told his students: "There is no geometric problem in which there is a need to study more than four derivatives."

Let us consider derivatives of area A and of Dirichlet's integral D at some minimal surface X which is not immersed, i.e. which has branch points. The first derivative of A is already a nontrivial matter, as $|X_u \wedge X_v|$ appears in the denominator when one differentiates A, and the computation of higher order variations of A might seem hopeless. On the other hand, the conformally parametrized regular extremals of A and D agree, and one even knows that $\inf_{\mathcal{C}(\Gamma)} A = \inf_{\mathcal{C}(\Gamma)} D$. This suggests that one might be able to study higher order variations of D. At first glance this might seem to be trivial since the integrand is quadratic; but Plateau's boundary condition is highly non-linear, and so the computation of higher derivatives of

D turns out to be quite complicated if one also varies X on the boundary. A first difficulty is that, beyond the order 3, higher order derivatives of D are not intrinsic, e.g. they are not multilinear forms on a tangent space of the manifold of surfaces spanning a given contour. A second difficulty, reflecting the first, is the great complexity of calculating variations of D beyond the second. Only the special form of the variations, employed in this book, together with the use of Cauchy's integral theorem and the residue formula, made it possible to succeed. Hence it is not clear how this method could be applied to other integrals than D, and so an application to H-surfaces or to minimal surfaces in a Riemannian manifold seems presently to be excluded.

The approach to branch points of minimal surfaces, presented in this book, was discovered and developed by A.J. Tromba, with minor revisions by S. Hildebrandt for presentation in this volume. The somewhat indirect concept of true and false branch points is replaced by the concept of *exceptional* and *non-exceptional* branch points, which is formulated in terms of the order n and the index m of a branch point:

$$\text{either} \quad m+1 \equiv 0 \bmod (n+1), \qquad \text{or} \quad m+1 \not\equiv 0 \bmod (n+1).$$

The advantage of the results in Chap. 8 is that they apply to C^∞-contours and not only to real analytic boundary curves, but since not all boundary branch points are excluded, they presently do not cover the results of Gulliver–Lesley and B. White.

Appendix
Non-exceptional Branch Points; The Vanishing of the Lth Derivative, L Even

We discuss how to demonstrate that a minimal surface X with a non-exceptional branch point at $w = 0$ of order n, L even cannot be a minimum, if we consider the generator $\tau := c \in w^{-(n+1)} + \overline{c}\,\overline{w}^{n-1} + \delta c w^{-r} + \overline{\delta}\,\overline{c}\,\overline{w}^{r}$. As we have observed, the ϵ^{L-1} term of the Lth derivative is zero.

Here we need a trick. Going to the next highest derivative gives us additional parameters to work with, allowing us to show that, with appropriate choices, the leading ϵ^{L-1} term of the $(L+1)^{\text{st}}$ derivative is negative. This implies

$$E^{(L)}(0) = O(\epsilon^{L}), \tag{A.1}$$

and this remains true if we change the choice for τ to

$$\tau := c\epsilon w^{-n-1} + \overline{c}\epsilon w^{n+1} + \delta c w^{-r} + \overline{\delta}\,\overline{c} w^{r} + \rho, \quad c \in \mathbb{C}. \tag{A.2}$$

We infer

$$2(m+1) = L(n+1) - (n+1-r), \quad L = \text{even},$$

and so $n+1-r$ is even, which implies that

$$1 \le r \le n-1. \tag{A.3}$$

Next we define a meromorphic $\phi_t(0)$, real on S^1, such that

$$\begin{aligned} \phi_t(0) &:= -i\mu c^2\epsilon^2 w^{-2n-1} - i\delta c^2\epsilon(n+1-r)w^{-n-1-r} + \gamma c^2 w^{-r} + \cdots, \\ \gamma &:= \lambda/(n+1-r), \end{aligned} \tag{A.4}$$

with an arbitrary $\lambda \in \mathbb{C}$. Then it follows

$$\begin{aligned} w\hat{Z}_{ttw}(0)\phi(0) + 2w\hat{Z}_{tw}(0) = {} & 2c^3\epsilon(i\lambda A_1 w^{-r} + i\rho\delta(2n+2-r)A, w^{-r}\cdots, \ldots) \\ & + 2c^3\epsilon^3(\mu^2 A_1 w^{-2n} + \cdots, \ldots). \end{aligned}$$

For simplicity, we shall assume, for the moment, that $\delta = 0$. This does not alter (A.1). This leads to the definition.

$$\phi_{tt}(0) := -2c^3\epsilon^3\mu^2 w^{-3n-1} - i\lambda\epsilon c^3 w^{-n-1-r} + \cdots, \tag{A.5}$$

A. Tromba, *A Theory of Branched Minimal Surfaces*,
Springer Monographs in Mathematics,
DOI 10.1007/978-3-642-25620-2,

and inductively to

$$D_t^\beta \phi(0) := c^{\beta+1}[\text{const}\,\epsilon^{\beta+1} w^{-(\beta+1)n-1} + \text{const}\,\epsilon^{\beta-1} w^{-(\beta-1)n-r-1} + \cdots]$$
$$\text{for } 1 \le \beta \le L/2 - 1. \tag{A.6}$$

Now we write the formula for $E^{(L+1)}(0)$ in a different order as

$$E^{(L+1)}(0) = I_0 + I_1 + I_2 + I_3 + I_4 + I_5 + I_6$$

with

$$I_0 := 4\,\text{Re}\int_{S^1} w[D_t^L \hat{Z}(0)]_w \tau\, dw;$$

$$I_1 := \sum_{M=s+2}^{L-1} \frac{4L!}{M!(L-M)!}\,\text{Re}\int_{S^1} w[D_t^M \hat{Z}(0)]_w g_{L-M}\, dw, \quad s := L/2;$$

$$I_2 := \frac{4L!}{(s+1)!(L-s-1)!}\,\text{Re}\int w\left[D_t^{s+1}\hat{Z}(0)\right]_w g_{L-s-1}\, dw;$$

$$I_3 := \frac{2L!}{s!s!}\,\text{Re}\int_{S^1} w[D_t^s \hat{Z}(0)]_w \cdot h_s\, dw, \quad \alpha + \beta = s;$$

$$I_4 := \frac{2L!}{\sigma!\sigma!}\,\text{Re}\int_{S^1} w[D_t^\sigma \hat{Z}(0)]_w \cdot h_\sigma\, dw, \quad \sigma = s - 1 = L/2 - 1,\ \beta = L - \sigma - \alpha;$$

$$I_5 := \sum_{M=2}^{s-2} \frac{2L!}{M!M!}\,\text{Re}\int_{S^1} w[D_t^M \hat{Z}(0)]_w \cdot h_M\, dw, \quad \beta = L - M - \alpha;$$

$$I_6 := \frac{4L!}{(L-1)!}\,\text{Re}\int_{S^1} w\hat{Z}_{tw}(0) \cdot \hat{X}_w D_t^{L-1}\phi(0)\, dw$$
$$+2\,\text{Re}\int_{S^1} w\hat{X}_w \cdot \hat{X}_w D_t^L \phi(0)\, dw.$$

The standard reasoning yields $I_0 = 0$, $I_6 = 0$, and the pole-removal process yields that g_{L-M} is holomorphic; thus also $I_1 = 0$.

Lemma A.1 *We have $I_5 = O(\epsilon^L)$. This leaves us with*

$$E^{(L+1)}(0) = I_2 + I_3 + I_4 + O(\epsilon^L). \tag{A.7}$$

Proof We begin by considering the contribution of the last complex component to I_5. Since $\alpha \le M$, we have in

$$w[D_t^M \hat{Z}(0)]_w \cdot h_M, \quad h_M = \sum_{\alpha=0}^{M} c_{\alpha\beta}^M \psi(M,\alpha)[D_t^\alpha \hat{Z}(0)]_w D_t^\beta \phi(0), \tag{A.8}$$

that $\beta = L - M - \alpha \ge L - 2(s-2) = L - 2(L/2 - 2) \ge 4$.

We will show that there is no pole associated with a term that has ϵ^γ, $\gamma \le L-1$, as a coefficient. We have

$$w[D_t^M \hat{Z}(0)]_w \cdot [D_t^\alpha \hat{Z}(0)]_w = \text{const}\,\epsilon^{M+\alpha} w^{1+2m-(\alpha+M)(n+1)} + \cdots.$$

In order to achieve a coefficient of order ϵ^γ, $\gamma \le L-1$, we must consider the contribution from the second term of (A.6),

$$\text{const}\,\epsilon^{\beta-1} w^{-(\beta-1)n-r-1}, \quad M+\alpha+\beta = L.$$

The order of the w-term will then be

$$\begin{aligned} &1+2m-(M+\alpha)(n+1)-(\beta-1)n-r-1 \\ &\quad = (2m+2)-(M+\alpha+\beta-1)(n+1)+(\beta-1)-r-2 \\ &\quad = r+(\beta-3)-r \ge 1; \end{aligned}$$

thus there is no pole.

The lowest w-powers associated to ϵ^{M-2} in $(D_t^M \hat{Z}(0))_w$ are of the order $m-(M-2)(n+1)$.

Considering the order of the largest pole in $D_t^\beta \phi(0)$ with coefficient $\epsilon^{\beta+1}$ and looking at the total contribution to a pole of order ϵ^{L-1} in (A.8), we obtain a term of the form

$$\text{const}\,\epsilon^{M-2+\alpha+(\beta+1)} w^{1+2m-(M-2)(n+1)-\alpha(n+1)-(\beta+1)n-1} = \text{const}\,\epsilon^{L-1} w^{r+\beta-1};$$

so again there is no pole.

What about the first two complex components? In the first case from above we get terms of the form ($j \le 2m-2n$):

$$\text{const}\,\epsilon^{M+\alpha+\beta-1}(A_j \cdot A_{2m-2n+1}) w^{1+2m-n-\alpha(n+1)-(\beta-1)n-r-1},$$

and

$$\begin{aligned} &1+2m-n-\alpha(n+1)-(\beta-1)n-r-1 \\ &\quad = 2+2m-n-\alpha(n+1)-\beta n+n-r-2 \\ &\quad = 2+2m-\alpha(n+1)-\beta(n+1)+\beta-r-2 \\ &\quad = 2+2m-(L-M)(n+1)+\beta-r-2 \\ &\quad = 2+2m-(L-1)(n+1)+(M-1)(n+1)+\beta-r-2 \\ &\quad = r+(M-1)(n+1)+\beta-r-2 \\ &\quad = (M-1)(n+1)+\beta-2 \ge (n+1)+2 = n+3 \end{aligned}$$

since $M \ge 2$ and $k \ge 4$.

Again, there is no pole, and similarly for the second case from above. This completes the proof of Lemma A.1. □

Lemma A.2 *We have*

$$I_3 = O(\epsilon^L) \tag{A.9}$$

with a real number $T \ge 0$.

Proof We have $n \geq 2$ because, by (A.3),

$$n - 1 - r \geq 0, \quad n - 1 - r = \text{even}, \ r \geq 1.$$

We begin by considering what comes from the first two complex components in the products

$$w[D_t^s \hat{Z}(0)]_w \cdot [D_t^s \hat{Z}(0)]_w \tau, \quad w[D_t^s \hat{Z}(0)]_w \cdot [D_t^{s-1} \hat{Z}(0)]_w \phi_t(0), \quad \ldots \tag{A.10}$$

$s = L/2$.

(i) First we have to understand $[D_t^s \hat{Z}(0)]_w$:

$$\begin{aligned}
[D_t^s \hat{Z}(0)]_w &= \left\{ 2H \operatorname{Re}\left[i \sum_{\alpha+\beta=s-1} \frac{(s-1)!}{\alpha!\beta!} w [D_t^\alpha \hat{Z}(0)]_w D_t^\beta \phi(0) \right] \right\}_w \\
&= \left\{ 2H \operatorname{Re}\left[i w (D_t^{s-1} \hat{Z}(0))_w \tau + i(s-1)(D_t^{s-2} \hat{Z}(0))_w \phi_t(0) + \cdots \right] \right\}_w \\
&= \left\{ 2H \operatorname{Re}\left[c^s \epsilon^s i^s T_1 R_m w^{\gamma_1} + c^s i^{s-1} \epsilon^{s-2} T_2 R_m w^{\gamma_2} + \cdots \right] \right\}_w
\end{aligned}$$

where T_1, T_2 are real constants with $T_2 > 0$, and

$$\gamma_1 := -\frac{1}{2}(n + 1 - r) < 0, \qquad \gamma_2 := \frac{1}{2}(3n + 1 - r) > 0.$$

Recall that

$$\{2H[\operatorname{Re}(a w^{-\nu})]\}_w = \nu \bar{a} w^{\nu - 1}.$$

Hence,

$$\begin{aligned}
&\Big\{ 2H \Big[\operatorname{Re}(c^s \epsilon^s i^s T_1 R_m w^{\gamma_1} + \cdots + c^s \epsilon^s i^s T_4 w^{-1} + \cdots \\
&\quad + c^s i^{s-1} \epsilon^{s-2} T_2 R_m w^{\gamma_2} + \cdots) \Big] \Big\}_w \\
&= \bar{c}^s \epsilon^s (-1)^s i^s T_1 \bar{R}_m (-\gamma_1) w^{-\gamma_1 - 1} + \cdots + (-1)^s \bar{c}^s \epsilon^s i^s T_4 + \cdots \\
&\quad + c^s i^{s-1} \epsilon^{s-2} T_2 R_m \gamma_2 w^{\gamma_2 - 1} + \cdots .
\end{aligned}$$

Renaming $(-1)^{s+1} \gamma_1 T_1$ as T_1, $(-1)^s T_4$ as T_4, and $\gamma_2 T_2 > 0$ as T_2, we obtain

$$\begin{aligned}
[D_t^s \hat{Z}(0)]_w &= \bar{c}^s \epsilon^s i^s T_1 \bar{R}_m w^{-\gamma_1 - 1} + \cdots + \bar{c}^s \epsilon^s i^s T_4 \\
&\quad + \cdots + c^s i^{s-1} \epsilon^{s-2} T_2 R_m w^{\gamma_2 - 1} + \cdots
\end{aligned} \tag{A.11}$$

whence

$$\begin{aligned}
w[D_t^s \hat{Z}(0)]_w \cdot [D_t^s \hat{Z}(0)]_w &= (i^{L-1} |c|^L \epsilon^{L-2} T_1 T_2 |R_m|^2 w^\gamma + \cdots \\
&\quad + 2 i^{L-1} |c|^L \epsilon^L T_2 T_4 R_m w^{\gamma_2} + \cdots) + O(\epsilon^{L-1})
\end{aligned}$$

with $\gamma := -\gamma_1 - 1 + \gamma_2 = \frac{1}{2}(n + 1 - r) + \frac{1}{2}(3n + 1 - r) - 1 = 2n - r$.

Multiplication by τ yields

$$\begin{aligned} w[D_t^s \hat{Z}(0)]_w \cdot [D_t^s \hat{Z}(0)]_w \tau = &(c|c|^L \epsilon^{L-1} i^{L-1} T_1 T_2 |R_m|^2 w^{\gamma-(n+1)} + \cdots \\ &+ 2c|c|^L \epsilon^{L-1} T_2 T_4 R_m w^{\gamma_2-(n+1)} + \cdots) \\ &+ O(\epsilon^L) \end{aligned}$$

where $\gamma - (n+1) = (2n-r) - (n+1) = (n-1) - r \geq 0$ and $\gamma_2 - (n+1) = \frac{1}{2}(n-1-r) \geq 0$.

Thus, we obtain

$$\int_{S^1} w[D_t^s \hat{Z}(0)]_w \cdot [D_t^s \hat{Z}(0)]_w \tau dw = O(\epsilon^L). \tag{A.12}$$

(ii) Next we claim that there is no contribution of order ϵ^{L-1} or lower which comes from any of the complex components of the terms in (A.10) which are indicated by $\ldots$, that is, from

$$w[D_t^s \hat{Z}(0)]_w [D_t^{s-\beta} \hat{Z}(0)]_w D_t^\beta \phi(0) \quad \text{for } \beta > 1. \tag{A.13}$$

Recall that $\phi_{tt}(0)$ is defined so that

$$w\hat{Z}_{ttw}(0)\tau + 2w\hat{Z}_{tw}(0)\phi_t(0) + w\hat{X}_w \phi_{tt}(0)$$

has no poles. As noted in (A.5),

$$\begin{aligned} \phi_{tt}(0) = &-ic^3\epsilon(n+1-r)\lambda w^{-n-1-r} - 2c^3\mu^2\epsilon^3 w^{-3n-1} \\ &+ \text{terms with lower order poles.} \end{aligned} \tag{A.14}$$

Using this we see that

$$w[D_t^s \hat{Z}(0)]_w \cdot [D_t^{s-2} \hat{Z}(0)]_w D_t^2 \phi(0)$$

has no pole associated with coefficients of order ϵ^L or lower, and similarly for all $\beta > 2$.

(iii) Now we investigate in the second term of (A.10) what contribution comes from the third complex component of the terms involved. This contribution, $\tilde{C}$, is

$$\begin{aligned} \tilde{C} = \int w \cdot [&\overline{c}^s \epsilon^s i^s T_1 \overline{R}_m w^{-\gamma_1 - 1} + \cdots + \overline{c}^s \epsilon^s i^s T_4 + \cdots \\ &+ c^s \epsilon^{s-2} i^{s-1} T_2 R_m w^{\gamma_2 - 1} + \cdots] \\ &\cdot [i^{s-1}\epsilon^{s-1}c^{s-1} R_m T_3 w^{\gamma_3} + \cdots] \cdot (-i\epsilon^2 \mu c^2 w^{-2n-1} + \gamma c^2 w^{-r} + \cdots) \end{aligned}$$

where $\gamma_3 := \frac{1}{2}(r+n+1)$ and $T_3 \geq 0$.

This leads to

$$\begin{aligned} \tilde{C} = \int [&\gamma |c|^L c i^{L-1} \epsilon^{L-1} T_1 T_3 |R_m|^2 w^{\gamma_3 - \gamma_1 - r} + \cdots \\ &+ \gamma c |c|^L i^{L-1} \epsilon^{L-1} T_4 R_m T_3 w^{\gamma_3 - r} + \cdots \\ &- c^{L+1} \epsilon^{L-1} i^{L-1} T_2 T_3 \mu R_m^2 w^{\gamma_2 + \gamma_3 - 2n - 1}] + \cdots + O(\epsilon^L) \end{aligned}$$

and

$$\gamma_3 - \gamma_1 - r = \frac{1}{2}(r+n+1) + \frac{1}{2}(n+1-r) - r = n - r + 1 > 0,$$
$$\gamma_2 + \gamma_3 - 2n - 1 = \frac{1}{2}(3n+1-r) - 1 + \frac{1}{2}(r+n+1) - 2n - 1 = 0,$$

and $\gamma_3 - r > 0$. Thus we obtain

$$\tilde{C} = O(\epsilon^L). \tag{A.15}$$

Now we have to study the contributions coming from the first complex components.

(iv) The first term of (A.8) will have a lowest w-power of the form

$$\epsilon^s \cdot \epsilon^s (A_j \cdot A_{2m-2n+1}) w^\gamma$$

for some $j \le 2m - 2n$ and

$$\begin{aligned}\gamma &:= 1 + 2m + n - s(n+1) - (n+1)\\ &= \frac{1}{2}\{2(2m+2) - 2n - L(n+1) + 2n - 4\}\\ &= \frac{1}{2}\{[(2m+2) - (L-1)(n+1)] + [(2m+2) - (n+1) - 4]\}\\ &= \frac{1}{2}\{r + (L-2)(n+1) + r - 4\}.\end{aligned}$$

Since $L \ge 6$ it follows that

$$\gamma \ge \frac{1}{2}\{2r + 4(n+1) - 4\} = r + 2n > 0.$$

(v) Similarly the second term of (A.10) is harmless, and what we have seen in (ii) also applies to the other terms (A.13) of (A.10).

Inspecting (i)–(v) we obtain the assertion of Lemma A.2. □

We now need to investigate I_4, which is defined as

$$I_4 = \frac{2L!}{\sigma!\sigma!} \operatorname{Re} \int_{S^1} w[D_t^\sigma \hat{Z}(0)]_w \cdot \left\{ \sum_{\alpha=0}^{\sigma} \frac{\sigma!}{\alpha!\beta!} \psi(\sigma,\alpha)[D_t^\alpha \hat{Z}(0)]_w D_t^\beta \phi(0) \right\} dw \tag{A.16}$$

with $\sigma = s - 1 = L/2 - 1$ and $\alpha + \beta + \sigma = L$ whence $2 \le \beta \le L/2 + 1$.

Lemma A.3 *The terms in* (A.16) *with* $\beta \ge 3$ *are of the order* $O(\epsilon^L)$.

Proof We need to show that the terms with $\beta \ge 3$ and coefficients ϵ^{L-1} have no poles.

(i) For example, if we consider the lowest order zero of the last complex component of $w[D_t^\sigma \hat{Z}(0)]_w$ with coefficient ϵ^σ, this is a term of the form

$$\text{const}\,\epsilon^\sigma w^{1+m-\sigma(n+1)}.$$

Multiplication by const $\epsilon^\alpha w^{m-\alpha(n+1)}$ and then by const $\epsilon^{\beta-1} w^{-(\beta-1)n-r-1}$ yields

$$\text{const}\,\epsilon^{L-1} w^\gamma$$

with

$$\begin{aligned}\gamma &= 1+2m-\sigma(n+1)-\alpha(n+1)-(\beta-1)n-(r+1)\\ &= \beta-3 \geq 0,\end{aligned}$$

i.e. there is no pole.

If, on the other hand, we consider the contribution of $\epsilon^{\beta+1} w^{-(\beta+1)n-1}$, and consider also the term with coefficient $\epsilon^{\alpha-2}$ in $[D_t^\alpha \hat{Z}(0)]_w$, that is

$$\text{const}\,\epsilon^{\alpha-2} w^{m-(\alpha-2)(n+1)},$$

the total product will again be of the form const $\epsilon^{L-1} w^{\gamma'}$ with

$$\begin{aligned}\gamma' &= 1+2m-\sigma(n+1)-(\alpha-2)(n+1)-(r+1)-(\beta+1)n-1\\ &= \beta-2>0,\end{aligned}$$

again there is no pole.

(ii) What about the first complex components? The worst terms are of the form

$$\text{const}(A_j \cdot A_{2m-2n+1}) w^{1+2m+n-\alpha(n+1)-(\beta+1)n-1}$$

and

$$\begin{aligned}&1+2m+n-\alpha(n+1)-(\beta+1)n-1\\ &\quad= 1+2m+n-(n+1)-\beta n-n-1\\ &\quad= 1+2m-\alpha(n+1)-\beta(n+1)+\beta-1\\ &\quad= (2+2m)-(\alpha+\beta)(n+1)+\beta-2\\ &\quad= (2+2m)-(L/2+1)(n+1)+\beta-2\\ &\quad\geq (L-1)(n+1)-(L/2+1)(n+1)+\beta-2\\ &\quad= (L/2-2)(n+1)+\beta-2>\beta-2\geq 0\end{aligned}$$

since $\alpha+\beta = L/2+1$.

This completes the proof of Lemma A.3. □

From (A.5) and Lemma A.3 we infer

$$I_4 = \frac{2L!}{\sigma!\sigma!}\,\text{Re}\int_{S^1} w[D_t^\sigma \hat{Z}(0)]_w \cdot [D_t^\sigma \hat{Z}(0)]_w \phi_{tt}(0)\,dw + O(\epsilon^L). \quad \text{(A.17)}$$

In the product $w[D_t^\sigma \hat{Z}(0)] \cdot [D_t^\sigma \hat{Z}(0)]_w$ we can ignore the contributions from the first complex components, and from the last one we obtain (as in the case $2m+2 = L(n+1)$, L odd) the contribution

$$c^{2\sigma} i^{2\sigma} \epsilon^{2\sigma} k^2 R_m^2 w^\gamma + \cdots + O(\epsilon^{2\sigma+1}), \quad \sigma = \frac{L}{2} - 1, \quad 2\sigma = L - 2,$$

with

$$k := (m-n)(m+1-2(n+1)) \cdots (m+1-\sigma(n+1)) > 0$$

and

$$\gamma := 1 + 2m - (L-2)(n+1) = n + r.$$

We obtain for the integrand on the right-hand side of (A.17) the expansion

$$-2c^{L+1} \epsilon^{L-1} i^{L-1} k^2 \lambda R_m^2 w^{-1} + \cdots + O(\epsilon^L).$$

Thus we infer that

$$I_4 = -\frac{2(L-1)!}{\sigma!\sigma!} \operatorname{Re} \int_{S^1} c^{L+1} \epsilon^{L-1} i^{L-1} k^2 \lambda R_m^2 \frac{dw}{w} + O(\epsilon^L). \tag{A.18}$$

We must now investigate I_2. The first term of I_2 is (omitting constants)

$$\int_{S^1} w[D_t^{L/2+1} \hat{Z}(0)]_w \cdot D^{L/2-1} \hat{Z}(0) \phi(0)\, dw.$$

The term $w D_t^{L/2-1} \hat{Z}(0)$ has a highest order pole of the form $k_1 \epsilon^{L/2} w^{-\frac{1}{2}(n+1-r)}$, $k_1 \neq 0$, arising from the generator ce/w^{n+1}.

The term $w[D_t^{L/2+1} \hat{Z}(0)]_w$ is of the form $k_2 \epsilon^{L/2-1} w^{\frac{1}{2}(n+1-r)-1}$ and therefore has a zero of order $\frac{1}{2}(n-1-r)$ yielding a contribution to I_2 of the form

$$\epsilon^{L-1} \int_{S^1} k_1 k_2 R_m^2 / w\, dw$$

but we know nothing about k_2. However, the other terms in I_2 do not contribute, as the pole terms in $D_t^\beta \phi$, $\beta > 0$, arising from $c\epsilon/w^{n+1}$ have orders too low to contribute. The lack of information about k_2 means that we also have no information about the sum $I_2 + I_4$, which could have a zero ϵ^{L-1} term, yielding absolutely nothing.

The trick, in this case, is to choose $\delta \neq 0$, $\rho \neq 0$. Then we see that the $(L+1)^{\text{st}}$ derivative is of the form $I_2 + I_4 + O(\epsilon^L)$. Suppose that in this sum the $\epsilon^{L-1}\rho\delta$ term is zero for all choices of $\rho\delta$. Then, if we choose $\lambda := -i\rho\delta(2n+2-r)$, the λ linear terms in the higher order derivatives $D_t^\beta \phi$, $\beta \geq 2$, no longer contribute to the ϵ^{L-1} terms in $I_2 + I_4$. However, this means that there is no cancellation due to these derivatives, implying that $k_2 \neq 0$ and also that $I_4 = O(\epsilon^L)$ as well. Hence, if $\kappa = k_2 k_2 \neq 0$, it follows that

$$E^{(L+1)}(0) = -\frac{2 \cdot (L-1)!}{s!s!} \operatorname{Re} \int_{S^1} \epsilon^{L-1} c^{L+1} i^{L-1} \kappa R_m^2 \frac{dw}{w} + O(\epsilon^L) \tag{A.19}$$

where $\kappa \neq 0$.

Then by an appropriate choice of c and $\epsilon > 0$ we can make $E^{(L+1)}(0)$ negative. Hence there is a real $\nu > 0$ such that

$$E^{(L+1)}(0) = -2 \cdot (L+1)!\nu\epsilon^{L-1} + O(\epsilon^L), \tag{A.20}$$

whereas

$$E^{(L)}(0) = O(\epsilon^L). \tag{A.21}$$

Now we want to prove Proposition 4.2 of Chap. 4, using (A.20) and (A.21). In addition we need the following auxiliary result to be verified later.

Lemma A.4 *For $\alpha = 2, 3, \ldots, L$ there are constants $b_\alpha \in \mathbb{R}$ such that*

$$E^{(L+\alpha)}(0) = b_\alpha \epsilon^{L-\alpha+1} + O(\epsilon^{L-\alpha+2}). \tag{A.22}$$

Proof of Proposition 4.2 Let us write $\hat{Z}(t,\epsilon)$ instead of $\hat{Z}(t)$ in order to express the dependence of $\hat{Z}$ on t and ϵ, and set

$$E(t,\epsilon) := D(\hat{Z}(t,\epsilon)) \tag{A.23}$$

and so $E(0,0) = D(\hat{X})$. Applying Taylor's theorem with respect to t and recalling that $E^{(j)}(0,\epsilon) = 0$ for $1 \le j \le L-1$, $0 < \epsilon \le \epsilon_0$ and some $\epsilon_0 > 0$ we obtain

$$E(t,\epsilon) = D(\hat{X}) + \sum_{\alpha=0}^{L} \frac{1}{(L+\alpha)!} E^{(L+\alpha)}(0,\epsilon) t^{L+\alpha} + R(t,\epsilon) \tag{A.24}$$

where the remainder $R(t,\epsilon)$ can be estimated by

$$|R(t,\epsilon)| \le M|t|^{2L+1} \quad \text{if } |t| \le t_0 \text{ and } 0 < \epsilon < \epsilon_0, \tag{A.25}$$

for some constant $M > 0$ and some sufficiently small $t_0 > 0$. Choosing $\epsilon_0 > 0$ sufficiently small, we may assume the following, taking (A.20)–(A.22) into account: There are positive numbers $\nu, a, c_2, \ldots, c_L$ such that for $0 < \epsilon \le \epsilon_0$ we have

$$\begin{gathered} \frac{1}{L!} E^{(L)}(0,\epsilon) \le a\epsilon^L, \qquad \frac{1}{(L+1)!} E^{(L+1)}(0,\epsilon) \le -\nu\epsilon^{L-1}, \\ \frac{1}{(L+\alpha)!} E^{(L+\alpha)}(0,\epsilon) \le c_\alpha \epsilon^{L-\alpha+1}. \end{gathered} \tag{A.26}$$

From (A.24)–(A.26) we infer for $0 < t < t_0$ and $0 < \epsilon < \epsilon_0$ that

$$E(t,\epsilon) \le D(\hat{X}) + (\epsilon a - \nu t)\epsilon^{L-1} t^L + \sum_{\alpha=2}^{L} c_\alpha \epsilon^{L-\alpha+1} t^{L+\alpha} + M t^{2L+1}.$$

Setting $t := 2a\epsilon\nu^{-1}$ and choosing $\epsilon^* := \min\{\epsilon_0, (2a)^{-1}\nu t_0\}$ we obtain for $0 < \epsilon < \epsilon^*$ that

$$\begin{aligned} (\epsilon a - \nu t)\epsilon^{L-1} t^L &= -a\epsilon^L t^L = -2^L a^{L+1} \nu^{-L} \epsilon^{2L} \\ &= -b\epsilon^{2L} \quad \text{with } b = 2^L \nu^{-L} a^{L+1} > 0 \end{aligned}$$

and

$$\sum_{\alpha=2}^{L} c_\alpha \epsilon^{L-\alpha+1} t^{L+\alpha} + M t^{2L+1} \le M^* \epsilon^{2L+1}$$

with

$$M^* := \sum_{\alpha=2}^{L} c_\alpha \left(\frac{2a}{\nu}\right)^{L+\alpha} + M \left(\frac{2a}{\nu}\right)^{2L+1}.$$

This yields

$$E(t,\epsilon) \le D(\hat{X}) + (M^*\epsilon - b)\epsilon^{2L} \quad \text{for } t = \frac{2a\epsilon}{\nu} \text{ and } 0 < \epsilon < \epsilon^*.$$

Choosing $\epsilon_\ell := \min\{\epsilon^*, (2M^*\ell)^{-1} b\}$ and $t_\ell := 2a\epsilon_\ell \nu^{-1}$ we obtain

$$\epsilon_\ell \to +0, \qquad t_\ell \to +0 \quad \text{and} \quad E(t_\ell, \epsilon_\ell) < D(\hat{X}).$$

Thus Proposition 4.2 of Chap. 4 is proved. □

Therefore the proof of Theorem 4.1 in Chap. 4 is complete as soon as we have verified Lemma A.4.

Before we do that let us mention that Proposition 4.2 of Chap. 4 can certainly not be derived from (A.20) and (A.21) alone as one sees by the following

Example The function

$$f(t) := t^2(t-\epsilon)^2 = \epsilon^2 t^2 - 2\epsilon t^3 + t^4$$

satisfies $f''(0) = 2\epsilon^2$ and $f'''(0) = -12\epsilon$, but still $t = 0$ is even a global minimizer for f. This shows the need of further information, e.g. on the higher order Taylor coefficients, in order to ensure that the minimal surface $\hat{X}$ is not a local minimizer.

Instead of Lemma A.4 we state a somewhat stronger result which immediately yields the desired result.

Lemma A.5 *Let L be even and $Q := L + 2k + 2$, $k = 0, 1, \ldots, L-1$. Then*

$$E^{(Q-1)}(0) = O(\epsilon^\mu) \quad \textit{and} \quad E^{(Q)}(0) = O(\epsilon^\mu) \quad \textit{for } \mu := L - (k+1). \tag{A.27}$$

Proof We argue only for those terms of the integrands that come from the last complex components, as the reasoning for the first is similar (as usual). Furthermore we prove the statement only for $E^{(Q)}(0)$ since the reasoning for $E^{(Q-1)}(0)$ is the same.

Recall that

$$E^{(Q)}(0) = \sum c_{\alpha\beta\gamma} \operatorname{Re} \int_{S^1} w[D_t^\alpha \hat{Z}(0)]_w \cdot [D_t^\gamma \hat{Z}(0)]_w D_t^\beta \phi(0)\, dw, \quad \gamma \le \alpha.$$

Consider only those terms of $D_t^\beta \phi(0)$ arising from the term $c^2 w^{-r}$ in the definition (A.4) of $\phi_t(0)$ since these will contribute the lowest order ϵ-terms.

For $\gamma + \beta < L/2 - 1$ the sum

$$\sum c_{\alpha\beta\gamma} w[D_t^{\gamma} \hat{Z}(0)]_w D_t^{\beta} \phi(0) \tag{A.28}$$

has no pole. We will only consider the terms with $\beta = 0, 1$ since for $\beta > 1$ the pole orders decrease.

Let us investigate the term

$$w[D_t^{Q/2-1} \hat{Z}(0)]_w \cdot [D_t^{Q/2-1} \hat{Z}(0)]_w \phi_t(0). \tag{A.29}$$

The term $[D_t^{Q/2-1} \hat{Z}(0)]w$ has no pole, and terms of order ϵ^{ν}, $\nu < L/2$, contribute only holomorphic terms to the third complex component. Thus, if we can show that any term in $[D_t^{Q/2-1} \hat{Z}(0)]_w \phi_t(0)$ of order ϵ^{ν} with $\nu < L/2 - (k+1)$ contributes no poles, then the minimal order μ of any ϵ-coefficient will be

$$\mu = L/2 + L/2 - (k+1) = L - (k+1),$$

as claimed. Now

$$\begin{aligned} &[D_t^{L/2+k} \hat{Z}(0)]_w \\ &\quad = \Big\{ 2H \operatorname{Re}\Big[(D_t^{L/2+k-1} \hat{Z}(0))_w \tau + (D_t^{L/2+k-2} \hat{Z}(0))_w \phi_t(0) + \cdots \Big] \Big\}_w . \end{aligned}$$

The key fact is that in order to obtain one additional contribution from $\phi_t(0)$ (no ϵ) we need to go down two derivatives. Thus, in general, our construction shows that in the formulation of $[D_t^{L/2+k} \hat{Z}(0)]_w$ dropping down 2ρ orders in the derivatives of $\hat{Z}$ yields a contribution of ρ additional $\phi_t(0)$-terms. Choose ρ so that $\nu + 2\rho = L/2 + k + 2$, implying that

$$\rho \geq k + 1.$$

Then the contribution of the last complex component to (A.29) is

$$\epsilon^{\nu} R_m w^{1+m-[L/2+(k-2\rho)](n+1)-\rho r - r}, \quad \nu \leq L/2 - (k+2).$$

The w-exponent is equal to

$$\begin{aligned} \gamma &:= \frac{1}{2}\{(2m+2) - (L-1)(n+1) + [2(2\rho - k) - 1](n+1) - 2\rho r - 2r\} \\ &= \frac{1}{2}\{r + (4\rho - 2k - 1)(n+1) - 2(\rho+1)r\}. \end{aligned}$$

But $2\rho \geq 2k + 2$; therefore

$$\gamma \geq \frac{1}{2}\{r + (2\rho+1)(n+1) - 2(\rho+1)r\} = \frac{1}{2}\{(2\rho+1)(n-r+1)\},$$

and so $\gamma > 0$.

For $\gamma = L/2 + s$, $0 \leq s < k$, the same argument shows that for $\mu \leq L/2 - (s+2)$ no pole-term with a coefficient ϵ^{μ} arises in (A.28). This completes the proof of the lemma. □

References

Alt, H.W.
1. Verzweigungspunkte von H-Flächen I. Math. Z. **127**, 333–362 (1972)
2. Verzweigungspunkte von H-Flächen II. Math. Ann. **201**, 33–55 (1973)

Alt, H.W., Tomi, F.
1. Regularity and finiteness of solutions to the free boundary problem for minimal surfaces. Math. Z. **189**, 227–237 (1985)

Beeson, M.
1. On interior branch points of minimal surfaces. Math. Z. **171**, 133–154 (1980)

Böhme, R., Tromba, A.
1. The number of solutions to the classical Plateau problem is generically finite. Bull. Am. Math. Soc. **83**, 1043–1044 (1977)
2. The index theorem for classical minimal surface. Ann. Math. **113**, 447–499 (1981)

Courant, R.
1. On a generalized form of Plateau's problem. Trans. Am. Math. Soc. **50**, 40–47 (1941)

Dierkes, U., Hildebrandt, S., Sauvigny, F.
1. Minimal Surfaces. Springer-Verlag (2010)

Dierkes, U., Hildebrandt, S., Tromba, A.
1. Regularity of Minimal Surfaces. Springer-Verlag (2010)
2. Global Analysis of Minimal Surfaces. Springer-Verlag (2010)

Douglas, J.
1. Solution to the problem of Plateau. Trans. Am. Math. Soc. **33**, 263–321 (1931)
2. On sided minimal surfaces with a given boundary. Trans. Amer. Math. Soc. **34**, 731–756 (1932)

Gulliver, R.
1. Regularity of minimizing surfaces of prescribed mean curvature. Ann. Math. **97**, 275–305 (1973)
2. Branched immersions of surfaces and reduction of topological type I. Math. Z. **145**, 267–288 (1975)

A. Tromba, *A Theory of Branched Minimal Surfaces*,
Springer Monographs in Mathematics,
DOI 10.1007/978-3-642-25620-2,

3. Branched immersions of surfaces and reduction of topological type II. Math. Ann. **230**, 25–48 (1977)
4. A minimal surface with an atypical boundary branch point. Differential Geometry. Pitman Monographs Surveys Pure Appl. Math. **52**, 211–228 (1991)
5. Existence of surfaces of prescribed mean curvature vector. Math. Z. **131**, 117–140 (1973)

Gulliver, R., Lesley, F.D.

1. On boundary branch points of minimizing surfaces. Arch. Ration. Mech. Anal. **52**, 20–25 (1973)

Gulliver, R., Osserman, R., Royden, H.L.

1. A theory of branched immersions of surfaces. Am. J. Math. **95**, 750–812 (1973)

Gulliver, R., Tomi, F.

1. On false branch points of incompressible branched immersions. Manuscr. Math. **63**, 293–302 (1989)

Jiang, Hai-Yi

1. Jiang, Hai-Yi A general version of the Morse–Sard theorem. J. Zheijiang Univ. SCI, 754–759 (2004)

Micaleff, M.J., White, B.

1. The structure of branch points in minimal surfaces and in pseudoholomorphic curves. Ann. Math. **139**, 35–85 (1994)

Morrey, C.B.

1. On the solutions of quasi-linear elliptic partial differential equations. Trans. Am. Math. Soc. **43**, 126–166 (1938)

Nitsche, J.C.C.

1. The boundary behavior of minimal surfaces: Kellog's theorem and branch point on the boundary. Invent. Math. **8**, 313–333 (1969)

Osserman, R.

1. Branched immersions of surfaces. In: Symposia Mathematica of Istituto Nazionale di Alta Matematica Roma **10**, 141–158. Academic Press, London (1972)
2. A proof of the regularity everywhere of the classical solution to Plateau's problem. Ann. Math. **91**, 550–569 (1970)

Radó, T.

1. The problem of least area and the problem of Plateau. Math. Z. **32** 763–796 (1930)
2. On the problem of Plateau. Ergebnisse der Math. Band **2**, Springer, Berlin (1933)

Schüffler, K.

1. Isoliertheit und Stabilität von Flächen konstanter mittlerer Krümmung. Manuscr. Math. **40**, 1–15 (1982)

Steffen, K., Wente, H.

1. The non-existence of branch points in solutions to certain classes of Plateau type variational problems. Math. Z. **163**, 211–238 (1978)

Tomi, F.
1. Tomi, F. False branch points revisited (to appear)

Tomi, F., Tromba, A.J.
1. On the structure of the set of curves bounding minimal surfaces of prescribed degeneracy. J. Reine Angew. Math. **316**, 31–43 (1980)

Tromba, A.J.
1. Intrinsic third derivatives for Plateau's problem and the Morse inequalities for disc minimal surfaces in $\mathbb{R}^3$. Calc. Var. Partial Differ. Eqn. **1**, 335–353 (1993)

Wienholtz, D.
1. Wienholtz, D. Der Ausschlußvon eigentlichen Verzweigungspunkten bei Minimalflächen vom Typ der Kreisscheibe. Diplomarbeit, Universität München. SFB 256, Universität Bonn, Lecture Notes No. **37** (1996)
2. Zum Ausschlußvon Randverzweigungspunkten bei Minimalflächen. Bonner Math. Schr. **298**, Mathematisches Institut der Universität Bonn, Bonn (1997)
3. A method to exclude branch points of minimal surfaces. Calc. Var. Partial Differ. Equ. **7**, 219–247 (1998)

White, B.
1. Classical area minimizing surfaces with real analytic boundaries. Acta Math. **179**, 295–305 (1997)

Printed by Publishers' Graphics LLC USA
MO20120409-187
2012